Essential Primary Mathematics

A145715

Essential Primary Mathematics

Caroline Rickard

Open University Press
McGraw-Hill Education
McGraw-Hill House
Shoppenhangers Road
Maidenhead
Berkshire
England
SL6 2QL

email: enquiries@openup.co.uk
world wide web: www.openup.co.uk

and Two Penn Plaza, New York, NY 10121-2289, USA

First published 2013

A catalogue record of this book is available from the British Library

ISBN-13: 978-0-335-24702-8
ISBN-10: 0-335-24702-4
eISBN: 978-0-335-24703-5

Library of Congress Cataloging-in-Publication Data
CIP data applied for

Typeset by Aptara Inc.
Printed and bound by CPI Group (UK) Ltd, Croydon, CR0 4YY

Praise for this book

"This book is an absolute must for every primary teacher. The perfect blend of subject knowledge, common misconceptions, pupil activities and self-assessment questions will support all those who are feeling slightly less than confident about teaching a mathematical topic. It will also be invaluable to experienced teachers and subject leaders who wish to think more deeply about how to teach mathematics effectively."

Sue Davis, Primary PGCE Course Leader and Lecturer in Mathematics Education, University of Leicester, UK

"This book has the conversational style of an excellent mentor and/or tutor of primary mathematics. It offers advice and guidance on how to be an effective teacher of mathematics whilst still drawing the reader's attention to the importance of developing good subject knowledge, and how this can be addressed. Mathematical concepts are explained with reference to their theoretical underpinning and are then set in the context of real learning opportunities that illustrate good pedagogy.

There is a real emphasis on teaching for learning, and this is most evident in the introductory chapter which provides a brief discussion of the big issues currently being debated in the field of primary mathematics.

The consistent format of the subject chapters supports the reader's ability to plan and teach a wide range of appropriate activities based on rich mathematics. These are all neatly illustrated by children's drawings which bring the book to life.

This is an all encompassing text for any student or teacher of mathematics and will feature on my highly recommended reading list."

Paula Stone, Senior Lecturer Primary Education (Mathematics), Canterbury Christ Church University, UK

"This book is ideal for student and practicing teachers alike. The user-friendly format such as the overview of contents at the beginning of each chapter and the highlighting of key misconceptions in each area, make it easy to locate relevant information. Each chapter evolves logically through subject knowledge and progression in learning for children. This book stands out from other texts I have used as there is an extremely helpful section at the end of each chapter which provides suggested classroom activities with associated learning objectives for each area of mathematics. As

a final year student, I only wish this book had been available to me at the beginning of my course!"

Shelley Rogers, Student Teacher, University of Chichester, UK

"This book approaches the teaching of primary mathematics with a clear ethos, which is explained in the first chapter and then pervades all the suggestions and discussions which follow. The author deals with issues such as turning children's misconceptions and 'mistakes' into learning opportunities, provoking the children into communicating their reasoning and differentiating lessons in ways that empower rather than categorise children. The author's experience of having taught and observed hundreds of mathematics lessons is distilled into the essence of primary mathematics teaching."

Dr Marcus Witt, Senior Lecturer in Primary and Early Years Mathematics Education, University of the West of England, UK

"The theory sections of the book are really detailed which helps to provide a secure knowledge base for teaching primary mathematics. I especially like the way that the book is laid out; it is really easy to navigate. I love how the common misconceptions are outlined and explained in boxes separate to the body of the text but are also re-listed at the end of a chapter so that you could revise the potential issues which may arise before you start to teach a particular topic. The activities are well organised and adaptable but it is useful to see which age range each activity is suggested for at a glance, alongside the learning objective."

Natalie Ridler, NQT

Contents

Acknowledgements

I would like to record my grateful thanks to the many colleagues, family and friends who have encouraged and supported me. I owe an especially big thank you to Zachary who has kindly obliged his Aunty Caroline and supplied a good deal of the lovely art work and many of the mathematical examples.

1

Introduction

Welcome! This book aims to support you, the student teacher or current practitioner, in teaching mathematics in the best possible way. Each chapter outlines the essential knowledge required and shares some simple activities designed to support children in developing as confident and enthusiastic mathematicians. It would be both impossible and unnecessary to cover every aspect of subject knowledge in minute detail, and it may well be that what I have viewed as 'essential' is sufficient for you. Should you find yourself wanting a greater level of detail, however, then texts such as Haylock's comprehensive *Mathematics Explained for Primary Teachers* (2010) would be a good source of further information.

Each chapter has an introductory box as we see here, designed to give an overview of the chapter content which is then explored in the section 'What the teacher and children need to know and understand'. Whilst you might be keen to dip straight into some of the mathematical topics in the other chapters, note the suggestion that taking the time to read the content of *this* chapter could be beneficial!

About this chapter

A quick read of this introductory chapter is recommended since it sets the scene for the content chapters to come. We reflect on:

- Classroom ethos and valuing misconceptions
- Mathematical communication and reasoning
- Problem solving and investigation
- Lesson organisation
- Differentiation and attainment
- Subject and curriculum knowledge
- Developing children as independent mathematicians

Classroom ethos and valuing misconceptions

If learners are going to be keen to engage fully in mathematical activity, there are two key conditions:

1. The activities on offer need to be interesting enough to encourage children to engage with them.
2. A positive atmosphere needs to pervade the classroom such that children are willing to have a go at whatever is on offer, safe in the knowledge that getting something wrong is no big deal.

Misconceptions so often lead to rather intelligent wrong answers and are a really good window onto someone's thinking. This helps us to assess what it is that the children already understand and to appreciate the next useful stage in their conceptual development. Sometimes the misconceptions occur because of over-generalisation; learning something and then applying it more generally than is appropriate, such as assuming a cylinder is a prism (see Chapter 10). It is not generally considered to be one as it does not only have plane (flat) faces, but any child suggesting it is a prism is quite possibly demonstrating good knowledge about the properties of prisms, and this can of course be checked by asking them. If they talk about the 'ends' being the same shape and size and being parallel to each other, then they clearly already know a lot about prisms. Their misconception is therefore cause for celebration; it has uncovered the extent of what they do know as well as what they do not!

Mistakes in mathematics are often categorised as either simple slips (like an arithmetical error when adding) or more fundamental misconceptions. A range of common slips and more conceptual difficulties are detailed in each chapter to support you in your teaching, presented in separate text boxes.

The general consensus seems to be that rather than seeking to avoid misconceptions forming, teachers should anticipate that they will and purposefully introduce material likely to elicit them. This means they can be openly shared and addressed. So how do we respond when children have wrong answers? Well, one of the things we may well do is praise the child for having a go. But there is of course no one stock approach to dealing with wrong answers. On occasions we might solicit other answers and, having collected a range of answers, invite discussion on which the children agree with and why. At other times we might ask the child to explain his or her thinking, and this may well result in the child coming to realise an error. Interestingly, Cockburn and Littler (2008) remind us that we should also ask children with right answers to explain their thinking, as this sometimes uncovers misunderstandings where the right answer has been achieved but using a spurious method!

Mathematical communication and reasoning

As we have just seen, discussion is a good vehicle through which to explore misconceptions, but to what extent does talking feature in most mathematics lessons? From what my students tell me, the amount of talk varies a great deal from one classroom to the next and there are certainly arguments from both ends of the spectrum. There are times when working quietly and alone is a great opportunity for children to marshal their thoughts, but the clarity of those thoughts tends to get tested in dialogue with others, so both are important. I would also urge you to plan regular opportunities for children to have to convince you and each other of something; this is related to mathematical proof and provides a powerful reasoning opportunity. Some of the activities include what can best be described as 'red herrings' – the type of statement or example most likely to provoke discussion and quite possibly healthy disagreement!

Learning mathematics is somewhat akin to learning a foreign language: there are lots of new terms to learn, as well as familiar words used in rather different ways. 'Face' is a familiar, everyday word, but means something quite different when discussing solid shapes. Taking evening classes to learn a smattering of Spanish, our class were encouraged to take ownership of the new words we were learning by trying them out with each other. Mathematics is really no different, so it is important that the children get to talk mathematics on a regular basis. This includes discussion between children as well as with the teacher.

Questioning children is a classic tool of any teacher's trade, and good questioning has the potential to generate mathematical discussion and to enhance learning. Hewitt (2009: 8–10) makes an interesting distinction between types of question, and promotes what he calls 'honest' questions. If I ask a child what 5 + 5 equals, I already know the answer, whereas if I ask a child how they worked out that 5 + 5 = 10, this is an honest question since I do not already know the answer. Try to increase the number of honest questions in your teaching.

Clearly being fluent in communicating mathematically extends beyond discussion to being able to create a more permanent record of one's mathematics, for example, by writing it down in a way that another mathematician will understand. Early attempts at recorded mathematics (something we might liken to emergent writing) may well be idiosyncratic, and children will need support to move gradually towards recognising more conventional notation. Ideally, much of this recording will take place on blank sheets of paper or in exercise books. This may seem an odd comment, but this affords children opportunities to write things down for themselves and to rehearse the associated recording skills in a way that completing a worksheet cannot achieve.

Printed materials such as worksheets clearly do have a place in the mathematics classroom however, and it is possible to use these creatively with a little thought. The easiest way to approach this is to consider whether you could enhance children's opportunities for reasoning or communication; simple changes have the power to turn a satisfactory lesson into a good or really good one. If I have a sheet of 'sums', do I need the children to start at the beginning and work through them in order? Thankfully not, as this can result in mindless mathematical activity, such as using the same standard algorithm over and over again. Instead, I might ask the children to take a good look and discuss with the person next to them which question they think is the easiest and why. The dialogue would then be extended to find out whether everyone picked the same calculation and to explore variety in the children's methods. What about a particularly tricky calculation – could the children choose one they think might be a bit tougher and see if they can solve it with a friend? This type of lesson would feel very different compared to children working individually and in silence, but has far greater potential for reasoning and communication and thus more likelihood of quality learning taking place. Note, however, that the children may not plough through the whole worksheet in the time permitted; the emphasis here needs to be on quality rather than quantity.

When thinking about classroom ethos and imagining my ideal mathematics classroom, I realised that I would want all children to have to think pretty hard in my mathematics lessons! Rich activities, particularly problems and investigations allowing different avenues of exploration, encourage the development of powerful reasoning skills, and may well prompt mathematical communication.

Problem solving and investigation

Problem solving is a key feature of mathematics, and problems (requiring solutions) and investigations (not necessarily seeking solutions) range from those which are purely mathematical to those which allow us to apply mathematics to real-life or imaginary contexts. Just as rich activities promote reasoning, good opportunities for problem solving come from really rich mathematical environments where children develop a curiosity about the worlds of number, shape and so on, and have the chance to explore these worlds in an interesting and supported way. Whilst I am yet to be convinced about people having particular learning styles, any classroom which offers a rich mix of activity including visual, auditory and kinaesthetic opportunities helps to keep the learning experience lively and fresh.

Open-ended problems and investigations work particularly successfully precisely because they can be explored at a variety of levels and solved in different ways, allowing all children to participate in the process. Approaches to problem solving and investigation vary not just from person to person, but also because the problems posed vary in style. It is not unusual to hear problem solving interpreted only in terms of working on word problems, with them being tackled by reading them carefully (no bad thing), highlighting key words (which can be problematic, for example when there are words like 'more' which are used in different ways) and identifying and using the necessary operation. It is treating the choice of operation as a given that is the biggest problem for me ... If you ask someone how much change they would get from £20 after buying four items at £1·99 each, there are a number of different ways the problem could be tackled, using a mix of operations (+, −, ×, ÷) in a variety of ways.

Rather than proposing a set sequence of procedures to go through in solving problems, the following list is more like a buffet! I would rather children are given opportunities to pick from an extensive repertoire of problem-solving skills to explore a rich variety of problems. One enhances the other; the more of a range of problems I experience, the more problem-solving strategies I develop, and the better able I am to solve future problems since I have more strategies to draw on! For example, I might:

- Draw a picture. Pictorial images can help lots of people to understand problems.
- Gather resources. Concrete resources, as opposed to pictures, might help to model the problem in question. Acting some problems out is a possibility.

- Make a table. Tabulating information presented in the problem can make it easier to spot patterns, and extending the table to include other values may present a solution. Working systematically can be useful, particularly with problems where doing so has the potential to exhaust all possibilities.
- Try a simpler example. This is particularly useful in investigations, where it may well help to clarify what the problem is all about. Asking yourself whether you have tackled any similar problems in the past can also be helpful.
- Working from known facts or seeking new information. Check the problem does not contain any redundant information which you should be ignoring.
- Talk to someone else. The different points of view combined may help both learners on their way; talking to yourself can be useful too!
- Guess! This suggestion may come as a bit of a shock, and in my experience sometimes goes down least well with the most able mathematicians. But 'gut feeling' can provide a way in and, if all else fails, guessing may be a good option! I accept that I may not hit the solution on my first attempt, but if it helps me to understand the problem better and leads me to a solution via trial and improvement, then a guess can be a good thing. *Better* guessing can involve making an estimation and starting from there; or eliminating impossibilities and starting with something that remains.
- Work backwards. Some problems lend themselves to starting at an endpoint and 'reversing'.
- Have a cup of tea! A break from the problem may rest the brain enough to make just the leap in thinking that is required.

Note, too, that certain personal attributes such as perseverance are also key to solving tricky mathematical problems; there will be some dead-ends along the way, but having the resilience to retrace your steps and try a different approach is vital. By exploring all these sorts of approaches (and you will probably have your own ideas that you could add), children will hopefully develop the flexibility to attempt problems of any kind; these potentially have longer-term benefits than the ability to solve school word problems.

Lesson organisation

Having perused a chapter to reinforce your knowledge about an area of mathematics, you will hopefully then want to try out some of the activities in your own classroom. This will necessitate making some professional decisions about how the activities can be used and adapted to support what it is that you want the children to learn. The activities suggested throughout Chapters 2–11 are not designed for use in any particular part of the lesson; incorporate them where they work best for you and have the courage to tweak them when they would work well in a different format. Some activities may be best used in the more independent portion of the lesson; others in the teaching input or starter, some in the plenary, but you know your children and are therefore in the best position to meet their needs through reflecting on how to package the elements of the lesson. This is something that relying on an impersonal scheme or somebody else's worksheet may well not achieve. As well as adapting activities and deciding when to use them and for how long, the planning stage will also include classroom organisation decisions such as how you will arrange the furniture and organise the children to maximise learning.

Finding a really good way in to a lesson is important as it sets the tone for the rest of the lesson and makes the children want to be there rather than still out at play! So what makes a good starter? In the vein of the Primary Framework (DfES, 2006), what you want to do is to warm up the children's brains and help them to feel prepared for the lesson ahead, but opinion seems to be divided as to whether the starter should link to the rest of the lesson. Wherever possible, I would advise that it should. Watching hundreds of lessons in my current role, the strongest lessons are those which focus on the same topic throughout, building a coherent whole. I often advocate planning the teaching input and any independent tasks before going back and planning the start of the lesson; by that stage you are more likely to know what the most relevant warm-up will be in preparation for what is to follow. Finally, of course, you need to think about how to draw the lesson to a close, avoiding a random 'show and tell' approach. If you want children to show their work, whose will you showcase and why? Think carefully about what you want this portion of the lesson to achieve.

Let us now think a little more about the teaching that is to take place in the lesson. I suspect that the references to a three-part lesson when the National Numeracy Strategy was first introduced (DfE, 1999) caused a little confusion for some. Years later I still have students who refer to this and who plan lessons which have a mental and oral starter, some independent tasks for the children and finally a plenary. What is conspicuously absent is any direct teaching. Students' lessons sometimes go from the starter straight into explaining to the children what they will be doing independently (often in their ability groups). This is *not* teaching. The focus here is on what children will do rather than what they will learn, whereas the effective teacher will craft a sequence of little snippets of teaching to allow that learning to take place. This could include techniques such as: asking challenging questions to generate discussion and then teasing out key learning points; exploring equipment or images designed to exemplify teaching points; modelling conventional mathematical recording; telling mathematical stories or drawing on contexts to bring the mathematics to life. Hopefully you will be able to add to this list, whilst making sure not to omit this crucial part of the lesson.

Turning our attention now to the independent portion of the lesson, we need to consider which children will work totally independently and who would benefit from some further input from you (or a teaching assistant). Here, too, the teaching element should not be overlooked; keeping children on task is not efficient use of your teaching time; instead ask yourself what more you can contribute to the children's learning. In preparing work for those children who will work without adult support you will need to consider whether this will be undertaken individually or collaboratively. If group collaboration is planned, groups of what size? Pairs perhaps? And will the children be able to choose whom to work with, or do you want to engineer the groups in some way? An effective teacher of mathematics varies all these things according to what the lesson is designed to achieve, so there is no easy answer to the organisation of this part of the lesson. Be careful, however, not to fall into the trap of doing things the same way every day. Take, for example, the large number of students who report classrooms where the children are grouped by ability and the teaching assistant always works with the lowest-ability group. This begs all sorts of questions: when do the children get specialist input from the teacher? How can the teacher be sure of each child's ability? Are these children going to get the opportunity to learn some mathematical independence at some point? You may have further reflections of your own – I hope you do!

Resourcing the lesson will be another thought on your mind at the planning stage. You will notice that many of the activities proposed in the book use minimal resources, and often incorporate everyday items. The use of easily available materials has many benefits, including the opportunity to make connections between the field of mathematics and the real world. Resourcing in this way is cheap but also minimises the amount of time needed to prepare lesson materials. A blank sheet of paper (or the child's maths book, or a whiteboard) can sometimes be pretty much all that is required. Should you find yourself teaching in such a way that you need to make and explain multiple worksheets every mathematics lesson, stop and take stock; for your own sanity (as well as the children's learning), consider trying some of the ideas presented in this book.

What you will notice when you delve into the chapters is that many of the proposed activities are incredibly simple and can be adapted for use within different areas of mathematics (and quite possibly beyond) and with learners of different ages and abilities. Some of the best mathematics lessons I have ever observed are based around simple activity; the quality of the lesson is rarely directly linked to the quantity of time spent planning and preparing resources! Having said that, you do as the teacher have certain responsibilities in preparing for teaching these good-quality lessons and this will become easier with time and experience. Suppose you are preparing a range of subtraction calculations for the children to choose from (see Chapter 3) and you want the children to become aware of different strategies they might employ; then you will need to think carefully about the calculations you will include to ensure that a range of approaches are likely to be exhibited. Too narrow a range, with similar types of calculation, and the likelihood of even thinking about different calculation strategies will be constrained. Compare, for example, 84 – 5 = ? and 84 – 59 = ? and think about how you would work them out, assuming you did not know the answers straight off. If you are not sure that you would work them out differently, perhaps try asking a few friends or family members; different methods are likely to emerge because of the numbers chosen for the calculations. The range of examples is also important in other areas of mathematics; if I repeatedly present a single triangle image, an equilateral triangle with a horizontal base (what is referred to as a 'prototypical' image), then the children in my class will form a very narrow concept of 'triangle'.

Differentiation and attainment

Once you dip into the activities, you will probably also notice that many of them involve elements of choice. People have often challenged me as to whether children make appropriate choices for themselves, voicing concern that they might stick to the easy options. I can honestly say that I have never found this to be a problem, partly because it so rarely persists if choice becomes the norm, but also because you will know your children well and individuals can be given a little nudge if you feel you want them to attempt something a little harder! This results in the removal of 'false ceilings', the sort that are created every time you give a particular ability group a particular worksheet. If I am deemed to be in a particular group, then I can only do that group's work and am therefore limited by the teacher's expectations of me, expectations which may or may not be right. Getting the level right for each individual on a particular day in a particular aspect of mathematics is very difficult, so you can help to remove that barrier by employing more open-ended tasks. I suspect you will be pleasantly surprised with what some of your children are capable of. Incidentally, a wonderful side-benefit to simple activities with elements of choice is that they are efficient to plan (single task, no need for multiple worksheets) and readily contribute to a good work–life balance.

Related to this is another distinct advantage to working with these more open-ended tasks, associated with quantity of work. We all know that output varies considerably from child to child: completing and recording three questions might be cause for celebration for Luke, whereas if Charlotte produces only ten answers, a little nag might be warranted, expecting her to produce a few more! Standard worksheets do not allow for this variation, and for the Lukes of this world finishing a piece of work becomes what feels like a depressingly unobtainable goal. Where children are creating their own mathematics to a greater extent, this goal is removed and replaced with a sense that everyone is expected to work hard. Note that it often also removes the need for extension tasks as the extension is in-built into many of the activities, having no defined end point.

While expectations regarding quantities of work may vary, expectations of what a child can achieve must always be high. The teacher's attitude must be that almost every child is capable of becoming a sound mathematician, leading on from the concerns about 'false ceilings' suggested above. When planning lessons, work from the premise that all children are on the same learning journey. I often use a ladder analogy: children all climbing the same ladder that lesson (a fractions ladder or whatever) but with some children currently on lower rungs. With the open-ended tasks, however, all children can potentially access the upper rungs since they are at least headed in the right direction. You will notice in many of the chapters that some of the example material is handwritten. I deliberated a while over whether this would look professional enough in a published book, but decided that it helps to make an important point. Good teaching is not about mathematics looking pretty, but about helping children to climb that ladder. In lessons where we write numbers and calculations up as an integral part of the lesson, those numbers and calculations can be so easily changed to match the responses of the children if some are finding work too easy or too hard. Plan the examples in advance, but be prepared to change your mind during the lesson in response to the children's needs.

Linked to this are the arguments that rattle on in the field of mathematics education regarding ability grouping and setting, popular in the UK to a far greater extent than in many other countries. Having taught a class of mixed age and ability for seven years and used lots of my open-ended activity ideas, I am confident that working with others of whatever ability is the key to success. This is the case, incidentally, for adult learners as well as children. I believe that the dialogue that occurs precisely because learners know different things, think in different ways, and so forth, results in learning for all. I recommend Boaler's (2009) intriguingly titled *The Elephant in the Classroom* should you wish to read more about this topic.

Subject and curriculum knowledge

Amount of lesson preparation will depend to a certain extent upon how comfortable you feel about your own subject knowledge; the safest way of ensuring that you are well prepared is to attempt the mathematics yourself to uncover any glitches or surprises that might otherwise catch you out. For example, if children are choosing pairs of numbers to divide (see Chapter 5) and someone might select 5670 ÷ 45, make sure that you have thought about tackling such a calculation. Some years ago I observed a lesson focusing on rules for divisibility, from numbers ending in 0 being divisible by 10, to some more obscure ones. The trouble was that the student had got a bit muddled and failed to realise that some of his rules did not work. From then on he started trying out the mathematics in advance! Each chapter has four self-test questions at the end to give you a brief opportunity to check aspects of your own knowledge.

In contrast to subject knowledge, curriculum knowledge gives me an idea of what I should be teaching, but more important than this is time spent thinking about how to translate one's own knowledge into something accessible by the children. This means breaking the knowledge into manageable chunks and deciding how to bring it to life in ways the children will understand. You will notice that some of the activities are introduced almost as if they are written for you, the teacher. In many ways they are! The encouragement to try them out is deliberate since this will help you to understand them so much better. This in turn will help you to appreciate the best ways to use them in your teaching to meet the needs of your class, including making some things simpler and other aspects more challenging.

Earlier I mentioned the ideal of approaching lesson planning with the intention of building a coherent whole. However, we would also want to consider the potential links between lessons on the same topic, between different areas of mathematics, and between mathematics and other areas of the curriculum or with real life. I like to think of learning as a big jigsaw with pieces gradually slotting into place, and a still powerful study (Askew *et al.*, 1997) suggests that the most effective numeracy teachers make very explicit connections between different areas of mathematics. This, or the lack of it, was brought into sharp focus for me a few years ago when a postgraduate student was teaching a lesson on coordinates where the children were plotting points and joining them to create pictures of things like yachts and teddy bears. Looking through her files, I discovered that the previous day's lesson had been on triangles and in conversation learned that the children had struggled to get their heads round categorising different types of triangle, including isosceles and scalene triangles. What a missed opportunity to combine the two topics and potentially strengthen the learning of both of them. Aim to make the most of potential links yourself.

Use of ICT (information and communication technologies) can also strengthen the children's learning in mathematics lessons, but its effectiveness relies on the knowledge of the practitioner. Your depth of understanding of the mathematics is essential in order to make appropriate decisions about the role of ICT in supporting the development of mathematical concepts.

In previous sections we met the idea that children benefit from being immersed in activities which provoke their mathematical curiosity, such as rich problems and open-ended investigations. In order to embrace what children offer, and such activities *will* result in things they notice and want to share, then your own subject knowledge has to be really secure. This means you can build confidently on the contributions and questions of the children, making connections with mathematics already learned and that which is still to come.

Developing children as independent mathematicians

Following on from thinking about your own ability to do the mathematics and to be sure that you have the right answers, you might also want to reflect on aspirations for the child to be in that same position. I once watched a young boy sitting at the back of a class carry out a calculation on his mini whiteboard and give the answer a large and confident tick. How wonderful, I thought, to have that level of certainty in his mathematics. I was therefore somewhat saddened and surprised by the student teacher's response to his showing of his board: rather than celebrating his confidence and success (it was, after all, the right answer), he was told to rub out his tick because she had not yet given the answer. The student and I discussed this at length after the lesson; hopefully she has gone on to respond in ways which encourage children to think about whether they are sure answers are right, and to be really glad when the children are able to justify their answers. Incidentally, it does not always have to be the child's own mathematics under discussion; as we will see in various chapters, engaging children in exploring whether answers are correct or statements true or false can potentially result in powerful learning. If we are aiming for mathematical independence, then this type of activity is a step in the right direction.

Conclusion

Many of the points made so far are mirrored in the Teachers' Standards effective from 1 September 2012 (http://www.education.gov.uk/schools/leadership/deployingstaff). The advice in this and later chapters will support you in the following aims:

1. Set high expectations which inspire, motivate and challenge pupils. (For example, by using rich tasks to engage the children and valuing opportunities to talk about the mathematics.)
2. Promote good progress and outcomes by pupils. (For example, by recognising that tasks with good potential for reasoning will build pupils' mathematical understanding.)
3. Demonstrate good subject and curriculum knowledge. (For example, by anticipating and addressing misconceptions.)

4. Plan and teach well-structured lessons. (For example, by planning explicit modelling of key points as part of a carefully crafted lesson.)
5. Adapt teaching to respond to the strengths and needs of all pupils. (For example, by using the open-ended investigations recommended as ideal for challenging pupils of *all* abilities.)

This book has not been linked explicitly to what children of a particular age might be expected to learn, but each activity does suggest year groups you might wish to consider using it with. This is only a rough guide and the age groups are quite broad in some cases! Many of the activities can be made far easier or more challenging simply by changing the numbers involved; for example, continuing a number pattern (see Chapter 6) could be 2, 4, 6, 8, ... or 0·2, 0·05, –0·1, –0·25, ... Using such activities is therefore advantageous because they can be adapted to meet the needs of children at different stages of development but who are in the same class.

In spite of the potential curriculum changes, much remains constant in mathematics: children will still need to learn to add, subtract, and so on; they will still exhibit the same misconceptions; and the open-ended activities will continue to contribute effectively to children developing mathematically. So as I said at the outset: welcome to my book and enjoy using it!

2

Counting and the number system

About this chapter

In this chapter we consider different features of the number system, including whole numbers, decimals and fractions. Aspects requiring particularly careful teaching are described and some common misconceptions aired. There are so many different resources to support learning about the number system that they are dealt with separately at the end of the chapter:

- Learning to count
  - Five principles associated with learning to count
  - Progression in counting
  - Different types of number
  - Order and position
- Writing numerical values
  - The place value of the counting numbers
  - Decimals
  - Fractions
  - The relationship between fractions and decimals
- Applications of the number system
  - Percentages
  - Ratio and proportion
- Resourcing related to different aspects of the number system
  - Counting materials
  - Structured counting materials
  - Written images of the number system
  - Money

What the teacher and children need to know and understand

Having a secure understanding of the number system and the ability to move around comfortably within it is an essential mathematical skill. Much of the content we meet here will prove to be key building blocks to support calculation (see Chapters 3, 4 and 5) and the content of other chapters (such as measures and data handling).

Learning to count

Initially this section focuses on children's early counting and their developing ability to verbalise the number system; later the relationship between this and how the numbers are written will be considered. After a quick look at different types of whole number, ability to both order and position numbers in relation to other numbers concludes the section.

Five principles associated with learning to count

Most children will start to engage in counting activities long before they start school, and these early experiences of number should be encouraged and celebrated. Whereas we appreciate that each counting number is a separate entity, in the very early stages children often hear just the string as a combined entity, 'onetwothreefour', etc.

To illustrate the principles involved in learning to count, I will tell you about my niece, Isabel, who has just turned 3 and who counts things on a regular basis. Her counting is often related to food: numbers of strawberries, raspberries, crisps – you get the picture! So can she count? Well, yes and no! Her appreciation of the five principles (Gelman and Gallistel, 1978) is gradually becoming more secure but we could not yet claim that she can count totally reliably or extensively. Let us analyse her current position:

- Purposeful counting revolves around the *cardinal principle* (or cardinality). Isabel demonstrates some appreciation of this as she understands that a 'how many . . . ?' question is answered by counting and stating the last number in the count. She also appreciates that she can ask for certain numbers of things, counting to check that she's got it right! When children just start to re-count in response to a 'so, how many are there?' question, this suggests they have not fully grasped cardinality; without this, progression to new concepts such as addition would be irrelevant.
- Wrong answers to 'how many . . . ?' questions often arise from issues with the *stable order principle*. She is gradually improving but the number names or 'tags' do still come out in the wrong order sometimes, particularly going awry above 'ten'.
- A common error associated with the *one-to-one principle* (where items are counted once but once only) is where the timing is out. Isabel used to point to things she

wanted to count but recite the number names more quickly or slowly, resulting in each item not getting its own 'tag'. Errors also occur, particularly where larger quantities are involved, when she loses track of which items she has counted, resulting in counting things more than once or missing them out altogether. The one-to-one principle benefits from strategies such as pointing to, moving or lining up the items to be counted. Eventually children develop strategies for counting items they cannot touch whilst still keeping track of the count, or counting large numbers of items by partitioning them into a number of smaller subgroups.

- The *order irrelevance principle* means that the counting process does not have to start with any particular item in order to find out how many things there are. Thus, with a developing grasp of cardinality, faced with a bowl of strawberries or whatever, it does not seem to concern Isabel where to start counting.
- I'm not sure about her understanding of the *abstraction principle*, though, a principle which suggests someone can count things they cannot see and more abstract things like ideas. I was conscious of this once with the daughter of a friend (Rebecca was slightly older than Isabel is now) when we were counting body parts (I have no idea why!). I asked her how many legs the cat had and she looked around but the cat was not in the room; after a moment's hesitation her face took on a real look of concentration and she came up with the right answer, having visualised the absent cat for the purpose of counting its legs. She seemed both surprised and pleased to have an answer to this counting question!

Progression in counting

In addition to the fundamental principles outlined above, there are further skills to be learned to ensure counting skills are used effectively in the longer term. Whilst early counting typically starts at 1, competent 'counters' can start at other numbers and count confidently both forwards and backwards, and this helps to underpin later work in addition and subtraction. Incidentally, a long-term goal has to be to progress beyond solely counting approaches to calculation, recognising that something like 27 + 5 will take us forwards in the number system but that rather than counting forwards in ones we might take a jump of three to 30 and then two more to get to 32. More on this in Chapter 3.

Common misconception

'The numbers go sixty-seven, sixty-eight, sixty-nine, sixty-ten...'

Counting across the tens boundaries can initially prove tricky for some children, unsurprising since the fact that ten follows nine is reinforced a great deal in early learning. Ability to cross other boundaries can also be challenging, such as knowing what happens after ninety-nine.

Once children have a good grasp of how the number system works, many will realise that counting forwards can go on forever, and even very young children can begin to appreciate the idea of infinity. If we were to keep counting backwards, of course, we would pass zero and enter the world of negative numbers, ideally exemplified through contexts such as temperature or metres below sea level.

Counting in other multiples and familiarity with patterns like 3, 6, 9, 12,...and 5, 10, 15, 20,...help to lay foundations for later work. A secure ability to count in tens from any number is a particularly useful skill since our number system operates in multiples of one, ten, a hundred, and so on; thus we get very distinct patterns, such as where the tens change but the units do not: 47, 57, 67,.... Another important pattern element relates to the fact that as we count in ones, we pass odd and even values alternately (see the relationship with division in Chapter 5).

Grasping conservation of number is another feature of a developing understanding of counting, recognising that even if I rearrange the items I have counted, there will still be the same number – as long as none are added or taken away of course!

'Subitising' refers to the ability to identify small numbers of things *without* counting. This also supports children's mathematical progression as familiarity with small quantities allows for more efficient working. My niece demonstrates this occasionally with very small numbers, as rather than counting to answer a 'how many...?' question she just gives an answer suggesting she can 'see' up to three or four items without having to count. Regular dice (Figure 2.1), and also dominoes, rely on something akin to subitising, where we become accustomed to the patterns of the dots and this familiarity enables us to recognise the values at a glance; tallying is a similar case in point.

Figure 2.1 Typical arrangements of dots

Different types of number

The counting numbers described so far are what are referred to as *cardinal* numbers, numbers being used to represent quantities, and in other chapters we will look more closely at prime numbers, square numbers, and so on. But in addition to cardinal numbers we also use numbers in a couple of other ways.

If I want to describe my position in a queue (I have visions of Chichester Post Office just before Christmas when you might start off being something like fourteenth in the queue!) or somebody's success in a race, then I will be using what are referred to as *ordinal* numbers. Integers (positive and negative whole numbers) are often also considered ordinal since position in the number system means we can make statements like '10 comes after 9 but before 11'. More care is needed with the negative integers!

Language specific to ordinal number is more challenging than for cardinal number, since we take some of the cardinal numbers and add an ending, but not always the same one, or we coin a new word altogether! Being aware of these complexities enables more careful planning of the support children may need:

- 'First', 'second' and, to a certain extent, 'third' benefit from being in common use but do not mirror the counting numbers closely so are terms that need learning in a way that some of the others do not.
- Other numbers which end in one, two and three are also affected by the linguistic differences; thus we get 'twenty-first', 'twenty-second', 'twenty-third'.
- 'Eleventh', 'twelfth' and 'thirteenth' are, however, exceptions to this rule, and the spelling change from the cardinal to the ordinal for 'twelve' to 'twelfth' should be noted.
- Whilst numbers ending in four, six, seven, eight and nine all sound the same when they get their 'th' ending, they are not necessarily spelt consistently ('eighth', 'ninth') and numbers ending in five also sound different ('fifth', 'sixty-fifth').
- 'Tenth' is relatively easy, but there are minor changes with the other multiples of ten giving us 'twentieth', 'thirtieth', etc.
- Irrespective of number, I can also refer to something as being 'last' and working backwards learn that I can also talk about second to last position, etc.

Dates make perhaps the most common use of ordinal number, and here much of your teaching emphasis will be on associating the right letters with the numbers, as per some of the issues detailed above.

Common misconception

'Today is the 21th of June'

Whilst many ordinal numbers end 'th', this is not one of them!

The final category of number children will come across is referred to as *nominal*; numbers being used as names or labels. If I want to catch the bus up to Midhurst, then I need a number 60; this purely tells me which route the bus is following and has nothing to do with number of buses (cardinality) or position (ordinality).

Order and position

Ability to order numbers is a crucial element of getting to grips with the number system, for example, knowing that 18 is a bigger number than 15. But focusing on the relative positions of numbers gives this an added dimension. I would want children to learn things like the fact that 15 is half-way between 10 and 20; that 18 is nearer to 20 than to 10. I would want them to be able to state the 'next-door' numbers: that 17 comes before 18, and 19 after it. This can profitably be linked to the use of a variety of number lines with children.

In the longer term, children need to develop their understanding of order and position to include non-integers, for example: where $2\frac{1}{3}$ would sit in relation to the whole numbers and other fractions about it; how to work out decimals that lie between whole numbers and other decimals; and the ability to position fractions and decimals in relation to each other, recognising equivalent values. Just as children begin to appreciate that the counting numbers go on forever, so too will they learn eventually that more values will always exist between numbers however close those values might be!

Writing numerical values

The place value of the counting numbers

Numbers have not been and are not always recorded using the Hindu-Arabic numerals in predominant use today. Take, for example, Roman numerals, where each of a range of letters represents a different quantity. This does not operate on the type of system now referred to as 'place value', where the position of the digit denotes its value so that we recognise that each of the 5s in 555 are worth different amounts. Order does, however, have a small part to play in Roman numerals; for example, X is worth ten, but IX is one less than ten and XI is one more than ten. It is possible to pick this out since the values are otherwise presented from largest to smallest. In such a system there is no need for zero to record absence of quantity, whereas our place value system requires something to show clearly that a column is empty, as in 203 where we have two hundreds and three units but no tens.

Whilst we economically reuse the same ten digits, 0 to 9, to create the rest of our number system, errors are prone to creep in until the child develops a really secure understanding of how the place value system operates. Developing this secure understanding involves raising the child's conscious awareness that value relates to position, and that the number of digits involved therefore helps to give us clues about the size of the number.

Common misconception

'Twenty-one is written 201'

This and similar errors are extremely common, and unsurprisingly so. When children are familiar with how the two separate parts, the tens and the units, are written, it is a reasonable assumption to join the two parts together in the way we see here. With lots of opportunities to discuss this, along with judicious use of resources, children will come to realise that 201 must be a hundreds number as it has three parts, and that if I want to write my chosen number, it will only be 21.

Activities (such as activity 2.7) can help to reinforce the role that zero plays in denoting empty columns, and I find that it helps to stress the nothingness of the zero in early teaching examples. For example, if I'm writing 70, it is seventy-nothing, not seventy-eight or seventy-three! Incidentally, I find myself referring to 'units' out of ingrained habit, whereas some authors and practitioners seem to prefer 'ones'. I don't think it matters, and ideally children will be exposed to a rich variety of mathematical language and become gradually comfortable whatever term is used. What is vital, however, is that children are regularly encouraged to articulate the numbers they have written; just being able to write down 5727 is not enough; I have to know how to read this, and children's difficulties will only come to light if speaking is part of your mathematics classroom.

When discussing place value we need to be aware that it can be interpreted in two subtly different ways; we might treat each digit as precisely that – a single digit which denotes a certain number of tens, units or hundreds; or we can focus on what is sometimes referred to a 'quantity value' where the 2 in 123 is treated as '20'. For further reflection on this you might want to access materials by Ian Thompson, for example 'Place Value?' (Thompson, 2009).

Both when learning to count and when discussing place value with children, a common difficulty is coping with the 'teen' numbers as they are not said in a way which naturally reflects the structure of their numerical form. For example, when writing or saying the word 'sixteen', the six part comes first and the ten after; in contrast, when written in numerals as '16', we see the 1 (for one ten) first followed by the 6 units. Once we get beyond 20, this becomes less problematic as the order of the numerals matches the way in which the number is said, with the tens before the units as in '26' (twenty-six), '34' (thirty-four), '48' (forty-eight), and so on. Incidentally, this fails to cause such problems in some languages (Chinese, for example) where language mirrors mathematical structure throughout and we get the equivalent of 'one ten seven' for 17. The words for the multiples of 10 are also more transparent in some languages than others, with English speakers expected to associate the ending 'ty' with 20, 30, etc.

Note that in English (and many other European languages) number names 'eleven' and 'twelve' and their associated number forms '11' and '12' also have to be learned, as their names offer no clues as to their size or structure. So what do we do to support children? Well, in my experience not worrying too much about the numbers

Common misconceptions

'Fifteen is written as 50 in numbers'

Muddling the 'teen' and 'ty' numbers is common as they sound very similar. Children therefore need to be encouraged to listen very carefully to learn to distinguish between the two.

'Sixteen is written as 61 in numbers'

For the reasons explored in this section, reversal of digits is also common.

11 to 19 is the best and most relaxed approach; work on numbers in the twenties to crack the relationship between how numbers are said and written, exploring the ideas of tens and units, later returning to the more difficult portion of the number system. Opportunities can and should be taken, however, to challenge children's understanding of place value (see, for example, activity 2.4).

Decimals

Children's earliest experience of decimals is likely to be in the context of something like money, where the decimal point is used to separate the whole pounds from the pennies or hundredths of a pound, as in £3·99. Note that the reading of decimals needs to be treated carefully, since we would read the monetary value slightly differently than 3·99 as a regular decimal. Activity 2.10 is a particularly good one for illustrating the role of the decimal point. I once tutored a girl whose addition of decimals homework was mostly right, but whose few wrong answers pointed to a fundamental misconception – that a number can have two decimal points. Stressing its role as a boundary marker between whole and less than whole may help in this respect.

Just as writing whole numbers involves learning that columns have particular values, so too do the positions beyond the decimal point, with tenths followed by hundredths, then thousandths, and so on. As with the linguistic similarities between 'teen' and 'ty', the endings here need to be stressed as well, or children will fail to distinguish between 'hundred' and 'hundredth', etc.

Common misconception

'Before and after the decimal point we have units and tens and hundreds'

To the left-hand side of the decimal point we do of course have units, tens and so on, and some learners seem to think that what happens to the right is a mirror image. This lack of understanding can be signalled by errors such as three-hundredths written 0·003 rather than 0·03.

Children will gradually come to realise that there are different types of decimal: those with a single digit after the decimal point such as 0·5; those with several digits such as 0·125 (one tenth, two hundredths and five thousandths, equivalent to $\frac{1}{8}$); and some decimals which just keep going. The most likely in the primary sector are recurring decimals, the sort you get when you divide 1 by 3 on a calculator, turning $\frac{1}{3}$ into a decimal of 0·33333333333333333 (to infinity), written as $0.\dot{3}$, or $\frac{1}{7}$ which is 0.142587 (with the two dots signalling the beginning and end of the recurring string). There are also decimals known as 'irrational' numbers: decimals which just keep going without any repetition to the pattern and which cannot be written as a fraction.

Where approximate answers are sufficient, values may be given to a specified number of 'significant figures'. I seem to recall my current house was bought for about £72,000 (oh, those were the days!) but the actual value could have been anything between £71,500 and £72,499 if the value has been given to just two significant figures. Note that to identify the digits which are significant we work from the left and rounding is taken into account; had the original cost been £71,846, think about what this would have been to one, two, three or four significant figures (see Self-Test Question 2). Significant figures operate in just the same way where decimals are involved, starting from the left irrespective of where the decimal point occurs. Sometimes, however, situations specify that answers should be given to a certain number of decimal places instead.

Fractions

When teaching fraction concepts, the main thing to keep in mind is that fractions can be used to describe parts of wholes (referred to sometimes as 'pizza fractions') as well as being quantity related ('Smarties fractions'). Unless your teaching targets both these elements, children will develop only a limited concept.

A nice way of illustrating both aspects is to use a real-life scenario such as decorating a cake. I want to divide my cake into however many equal pieces (say, fifths), and I want to ensure that I decorate it with, say, chocolate buttons, in a way that no one argues about how many buttons they get (Figure 2.2). Multiples of five will work well for fifths of course, helping to demonstrate very clearly the relationship between fractions and multiplication and division.

Figure 2.2 Cake illustrating different types of fraction

Just as the language of ordinal numbers has the potential to give us a few headaches, fraction terminology proves a similar challenge and shares some but not all features. Just as I can be third in a queue, I can also use 'third' as a fraction, describing a particular size slice of cake or whatever. For two or four parts, however, I require new vocabulary: 'half' and 'quarter' (though, incidentally, quarters are sometimes referred to as 'fourths' for example, in America). The act of dividing something up according to fractions is also linguistically interesting; we can halve and quarter things, but do not typically use other fractions as verbs.

Early work on fractions will focus on language without the symbolic notation, but once children start learning about the notation of fractions, you will find that many are already familiar with $\frac{1}{2}$ as a type of shorthand for writing 'half'; the teaching role is therefore to support the children in appreciating what the two numbers tell us (gradually progressing to referring to the top number as the 'numerator' and the bottom number as the 'denominator'). Rather than sticking just to small numbers of parts, I advise giving lots of extreme examples in your teaching. If there are 28 children in your class, then imagining a cake cut into twenty-eighths helps to illustrate that these will be smaller parts than just cutting the same cake in half. If the whole school wanted a slice, goodness, how small those slices would have to be! Although the practicality of cutting very small slivers of cake would make it tricky to ensure they were all the same size, some children will need it to be made quite explicit that fractions imply equal parts. I have certainly experienced even Year 5 children saying, 'But you didn't tell me they had to be equal!'

But just appreciating what the denominator tells us is not enough; if children fail to pay attention to the numerator, this can also cause problems. Too often children meet predominantly unit fractions (the one-over-whatever type) and this adds to the problem as it is then tempting to stop paying attention to the numerator. Food again provides a nice way of exemplifying key features. Picture a cake cut into eighths. If I'm feeling exceptionally greedy then I might munch through $\frac{7}{8}$ before stopping, thinking I really ought to leave the last slice for someone else. Equally, it could be that I am not a big fan of chocolate cake, so a single slice is sufficient, leaving plenty for everybody else. This sense of almost eating the whole cake or the whole bag of sweets or alternatively very little/very few is important as it draws the children's attention to the role of the numerator. In fact, the more you can give that top number a role the better; that is how we know how many parts to eat, or to colour, or give away or whatever. The other point for discussion is to draw attention to the complementary fractions: if I enjoy just a quarter of a bag of sweets, three-quarters are left for other people, and so on.

Although comparing fractions mathematically is straightforward enough, with $\frac{5}{8}$ being greater than $\frac{1}{2}$, a note of caution should perhaps be sounded. In real life it does depend on what it is a fraction of.... Things are straightforward as long as we are comparing the same type and size of item, whereas half a fairy cake and half a Victoria sponge are going to feel a bit different!

All the fractions we have considered so far have essentially been less than wholes, but children also need to work with examples involving both wholes and parts. Think about opportunities to discuss fractions like $\frac{16}{10}$, giving a variety of contexts to focus the discussion. If this were pizza, what does it tell us? Well, that the slice size may well

Common misconceptions

'99 is a big number so ninety-ninths must be huge'

To a certain extent the world of fractions is counter-intuitive; children need to appreciate that big numbers can mean small fractions. However, understanding the relationship between numerator and denominator is vital to fully appreciate the size of the fraction.

'I know eighths are smaller than halves so $\frac{5}{8}$ must be less than $\frac{1}{2}$'

A bag with eight sweets would help to illustrate the erroneous thinking here, since eating half would only involve eating four of the eight sweets.

be tenths (a pizza cut into 10 slices) and that I was probably quite hungry as I seem to have eaten more than a whole one. This may lead to exploration of the different ways in which this information can be recorded, either $\frac{16}{10}$ as we see here (an improper or top-heavy fraction) or $1\frac{6}{10}$.

The relationships between fractions and decimals

Decimals and fractions essentially help us to do the same thing, to communicate with other mathematicians about parts of numbers, and they feature heavily in everyday life. Children will need lots of opportunities to familiarise themselves with the relationships between fractions and decimals, and versions of games such as snap and dominoes and online games support this process, with facts like $\frac{1}{2} = 0{\cdot}5$ quickly becoming known once the games have been played a few times.

Counting activities need not be precluded at this stage; familiarity with the patterns created using fractions and decimals is as useful as those generated by whole numbers. Counting comfortably in halves or quarters suggests a child is familiar with the order in which such numbers occur, but children also need to think about the positioning of such numbers on a number line. This gives a somewhat different impression of fractions and decimals, progressing children's understanding of the number system, encouraging them to recognise the smaller building blocks offered by fractions and decimals rather than only whole numbers, multiples of 10, and so on.

Common misconception

'This decimal number pattern goes 0·5, 0·6, 0·7, 0·8, 0·9, 0·10'

Just as we might see misconceptions regarding what comes next with younger children and whole numbers, there can also be mistakes made with regard to decimals.

Applications of the number system

Percentages

Percentages are a mathematical invention drawing on what we already know about the number system to describe everyday situations in relation to a notional 100% or total amount. 'Per cent' means 'out of one hundred', thus a value of 30% could relate to thirty out of one hundred people responding in a particular way in a survey, or any other situation where we want to describe a proportion of a total. In this type of scenario we are not going to get statements relating to anything over 100%, since this represents the full amount and we are describing a proportion of it, but as we will see later, this is not always the case. Survey data, real or imagined, is ideal as a way of illustrating percentages in the early stages, particularly where we can imagine that 100 people were asked. Another scenario I have seen presented in schools is to plan the design of a garden on a 100 square grid where, for instance, 20% should be pond, 50% grass, and so on.

Of course, in reality, the numbers involved may well not be a multiple of 100. Were 9 out of a group of 25 children big fans of a certain pop star, this too could be described in percentages, with 36 (9 × 4) out of 100 (25 × 4) being equivalent and giving us a value of 36%. Numbers are not always as convenient as this, however. To turn a statement like 'of 28 pupils, 23 brought a packed lunch to school' into percentages, we'd quite likely use a calculator to divide 23 by 28 to turn what is essentially a fraction into a decimal (23 ÷ 28 = 0·821 rounded to three decimal places), and then multiply by a hundred to move from the decimal to a percentage, giving us approximately 82%. Note that occasionally such a calculation is not required: had it been 'of 28 pupils, 21 had brought a packed lunch to school' it is possible to simplify the fraction to $\frac{3}{4}$, which should allow many children to identify this as 75%. Along with quarters (multiples of 25%) we might expect known facts to include tenths (multiples of 10%), fifths (multiples of 20%) and correspondingly twentieths (multiples of 5%), and of course the fact that one-half is 50%.

> **Common misconception**
>
> **'You can never have more than 100%'**
>
> This is a common misconception, and not just amongst children. This is certainly no surprise, since we do not get values of more that 100% in *some* situations and there is therefore considerable room for confusion. If I claim that 150% of my class love PE, this is clearly nonsense; even if they *all* say they love PE, this will only ever reach 100% as this constitutes the whole class.
>
> So when do we get percentages above 100? Well, imagine the PE-loving class had only six Y2 children (tiny village school), when suddenly another nine 7-year-olds move into the area, increasing the number of Y2 children from six to fifteen. This is what an increase of 150% would look like: 100% (another six) plus 50% (another three) giving a total of nine extra children.

One of the key words in the illustration above is that of 'increase', and both increases and decreases are commonly exemplified through problems involving changes in price. Good news in the sales (Figure 2.3); not so good news with things like value added tax (VAT) increasing from 17½% to 20%!

Figure 2.3 A sale is on!

Slight caution is called for when thinking about the result of a certain percentage increase followed by a decrease of the same amount (or vice versa). It can be tempting to think that you will get back to your starting amount, whereas in reality you will not. Test your understanding of this at the end of the chapter.

As a final note, rather than seeing percentages as an isolated topic, it is worth bearing in mind the interrelationships with topics already covered and still to come, notably probability which draws upon percentages, fractions and decimals to describe the likelihood of something happening.

Ratio and proportion

Although ratio and proportion are combined here in the same section, and are definitely related in that they play similar roles, they are subtly different and it is this subtle distinction which we need to be clear about in our teaching. I often introduce the idea of ratio and proportion using a packet of fruit gums, explaining that I really do not like the orange ones and imagining, pessimistically, that when I open the packet I will probably find lots of this flavour. *Proportion* typically uses the language 'out of' and focuses attention on how many out of the whole (akin to percentages). For example, 15 orange sweets out of a packet of just 20 sweets would be pretty unlucky for me (75% the flavour I least like). Proportions are often displayed as fractions, $\frac{15}{20}$, and in this example can be simplified to $\frac{3}{4}$ of the packet being orange.

Ratio, in contrast, pits the orange sweets against the other colours we find in the packet, and to identify the total number of sweets we would have to add the two numbers.

So if I end up with a ratio of 2 orange sweets to 18 of other flavours then I got lucky with this packet of 20 fruit gums (2 + 18). Incidentally, the ratio here can also be simplified, giving a ratio of 1 : 9.

Ability to adapt recipes for more or fewer people requires a sound understanding of ratio, with multiplication key. Should the changes made not maintain the ratios involved, it is quite possible that whatever is being made will end up too sloppy or crumbly. A recipe might demand, amongst other things, a couple of eggs, 100 g of flour and 50 g of butter. Any change to the numbers of eggs has to be reflected in a scaling up or down of the other ingredients. So if I double the number of eggs (× 2) I double the other ingredients; if I halve the recipe and use only one egg ($\times \frac{1}{2}$) then I only need 50 g of flour ($100\text{ g} \times \frac{1}{2}$) and 25 g of butter ($50\text{ g} \times \frac{1}{2}$). What if I wanted to use three eggs? How much flour and butter would I need now? Here the multiplier is not as obvious, but we need to know what it is if we are going to scale up the quantities of flour and butter whilst ensuring we keep the ratios the same. Three eggs (the original two eggs plus one more) is $1\frac{1}{2}$ times the original number, so each of the other ingredients will need to be multiplied by $1\frac{1}{2}$ too.

Common misconception

'To make my recipe for more people I need to add the same amount to every ingredient'

The issue here is that the learner is thinking about ratio in terms of addition and not multiplication as it should be. In the earlier example, if we added 100 g of flour and 100 g of butter, the relationship between the quantities of the two ingredients would be altered and might not have the desired effect!

Children will benefit from opportunities to explore ratio in different contexts (see, for example, activity 2.18), and this can involve, for example, numbers, mass, capacity or length.

Resourcing related to different aspects of the number system

The aspects of the number system discussed so far can be effectively supported by thinking carefully about which resources will best support the children's understanding. There's so much to choose from that selecting the right tools for the job is important, hence the belief that resourcing deserves a section all of its own. Remember that effective use of resources is not the long-term goal. We want children to be able to access a great deal of mathematics without having to resort to the use of resources eventually, but the early concrete experiences lay the foundations for this, in particular helping to build useful models of the number system in the mind.

Counting materials

As pretty much anything can be counted, early counting materials should ideally relate to the interests of the children. For example, an autumn walk might prompt the gathering of some pretty leaves which then form the basis of some of the mathematics lessons and which will likely include counting. But progress in counting will depend on gradually introducing greater structure to the counting and related experiences.

Structured counting materials

Multilink is essentially an unstructured counting resource, but has the advantage that cubes can be joined together to make fives or tens or whatever to give children a greater sense of the structure and size of numbers. This more structured approach for working with numbers beyond 10 is particularly important for an appreciation of why numbers are written in a certain way. But making lots of sticks of 10 using Multilink would be onerous, so you might prefer to use base 10 apparatus with your class where the units, tens, hundreds and thousands are ready made (Figure 2.4). I have also seen some quite creative alternatives on the market, such as sweets (cardboard images in singles, tens, and jars of 100) but other contexts can also be used (see activity 2.5).

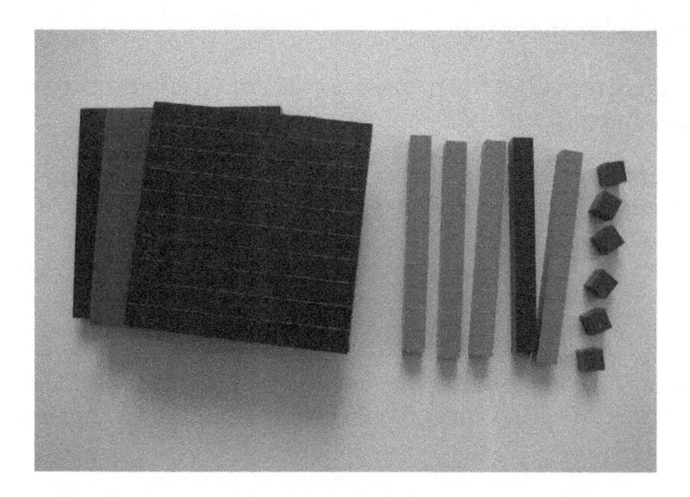

Figure 2.4 The number 356 shown using base 10 apparatus

Bead strings play a similar role since they are typically colour-coded in tens but the model is more akin to a number line image and may help to bridge the gap between physical representations of number (57 literally being shown by 57 beads or similar) and number lines representing the number system slightly more abstractly.

Numicon, a resource popular in many schools, offers a physical representation of numbers to 10, structured in such a way that the odd and even nature of each number is clearly demonstrated. Developing a feel for this is particularly useful in respect of combining and partitioning numbers, as children can visualise the constituent parts, for example, the fact that 7 is made up of a 4 and a 3 (Figure 2.5).

Figure 2.5 The number 7 partitioned using Numicon

Finally, resources such as dice and dominoes represent small quantities in an iconic way and can be useful in the mathematics classroom, as can many other everyday items!

Written images of the number system

All the resources we have looked at so far have one thing in common: they were all materials which represented numbers of things as opposed to how numbers are written. Where resources represent numbers of things, a sense of size is retained. For example, six dots on a dice will look like more than two dots on a dice; 356 represented using base 10 apparatus will physically correspond to having the equivalent of 356 little cubes and will take up more space than 256 would. When we get to resources which help with how numbers are written, however, it is not quite as easy to retain that sense of size. Take number fans, for example; to show 356 or any other three-digit number will require three 'petals' each time but will give no sense of the size of the number (Figure 2.6).

These, and resources like digit cards, certainly have a place in mathematics education, but the children will benefit from also using resources which help them to appreciate the relative sizes of the values involved.

Place value 'arrows' go part way to addressing this as the constituent parts do at least remind the user of the values represented by the tens and hundreds digits by showing the digits with their zeros (see activity 2.7). This essentially demonstrates the units, tens and hundreds as building blocks to be combined to make different size numbers. Being able to make and take apart such numbers is great for reinforcing the structure of the number system, and I particularly like the fact that the arrows come with colour-matched dice.

Figure 2.6 Number fans represent numbers using digits but offer no sense of quantity

Gattegno charts (Faux, 1998) perform a similar role since these too illustrate the building blocks and are particularly useful for dealing with decimals as well as whole numbers. In activity 2.8 a Gattegno chart is used alongside calculators to reinforce the structure of the number system to the right of the decimal point.

A key resource to support understanding of the number system (and therefore calculation) is the number line. We need to be aware, however, that there are different types and will want to make judicious choices as to which version best suits our purpose on different occasions. The earliest type of number line, sometimes referred to as a 'number track', is particularly useful for depicting the order of the whole numbers; each space on the number track has its own number. Numbered carpet tiles laid in order are a good example of the number system in this discrete form. In contrast, a true number line could be considered a continuous depiction of the number system, with the whole numbers evenly spaced and the potential for other numbers in the gaps. Because of this possibility for non-integers there are strong links to both measurement and data handling. As well as a range of positive values, number lines can also illustrate negative numbers, and are more likely to include zero. Washing line style number lines can be used in both ways and are commonly used for topics such as probability.

Whilst early number line work will often depict all whole numbers within a given range, children will gradually benefit from exposure to examples which show only some numbers, with multiples of 10 being particularly significant since many approaches to calculation benefit from a secure understanding of these 'markers'. Eventually this leads to something referred to as an 'empty number line' (ENL) being used more flexibly than a predetermined image of the number system. Here children move back and forth in the number system (either mentally or jotting such a line on paper), choosing their own markers as relevant to the calculation being attempted.

Like the discrete number track, a 100 square has each space labelled with a particular number. For a while it was fashionable to start at 0, but this results in finishing at 99, and I would much prefer my hundredth square labelled '100' and the first square '1'! Whilst 100 squares can be used in similar ways to number tracks, they have the added advantage of stressing the multiples of 10 as each row consists of ten squares and finishes explicitly with a multiple of 10, a benefit which is lost if you label the first square '0'! Either way, reading down the columns, each unit digit remains the same, helping to illustrate counting patterns such as 7, 17, 27, 37…

You may have come across Numdrums (Figure 2.7). These achieve similar aims to the 100 square (such as the patterns of the units), but do so in a continuous fashion with the numbers spiralling up around a tube, as opposed to 100 squares which typically work from top left. These can particularly help children who get to the end of a row on a 100 square and struggle to know which number comes next.

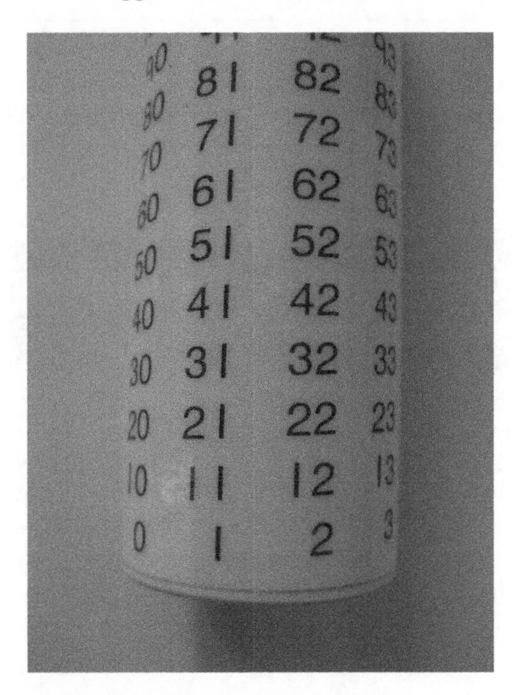

Figure 2.7 Part view of a Numdrum

Of course children will meet other depictions of the number system, such as the numbered squares on a snakes and ladder board. Relish such opportunities – becoming really familiar and comfortable with the number system is vital!

Money

Finally, let us think about money. In many ways, coins and notes belonged in an earlier section as they represent quantity, with each coin or note having a particular value. But in contrast to the other counting resources, with money it is much harder to establish that sense of size. That a tiny little 5p coin is worth more than a considerably larger 2p is bound to be confusing!

Common misconception

'I'm really rich because I've got lots of coins'

Treating coins as straightforward items to be counted in ones is a common misconception amongst young children, and incidentally, counting any coins other than pennies in ones should be avoided. Experience over time will teach children that a single pound coin is actually more valuable than quite a number of other options!

Once children have cracked the idea of value, however, money does give another resource we can use to model the structure of the number system. In pennies we might focus on combinations of 10p and 1p coins to mirror tens and units, or might use a combination of pounds and pence to exemplify decimals.

Activities for counting and understanding of the number system

Many of the following activities can be adapted for use with children of different ages and abilities and to target different aspects of the number system. When thinking about how you will use the activities (for example, which parts of your lessons they will prove most useful in), ensure that you build in opportunities for children to articulate their work. Being able to write mathematics is great, but we want children to be able to read what they have written, too!

Pupil activity 2.1

All sorts of number
Y1 to Y6

Learning objective: To begin to appreciate the many different categories numbers either belong to or are excluded from.

This is a really super activity for looking at different aspects of the number system. It involves choosing a number (completely personal choice) and then comparing that number with those chosen by your friends. Guessing is not the aim of the game, and children will need to be told to talk openly about their number with their friends, particularly looking for similarities and differences.

Let's take an example, say, 45 and 24. Depending on the age of the children, dialogue might identify the following:

- We haven't chosen the same number.
- 45 is bigger than 24.
- We have both chosen two two-digit numbers.
- Both numbers have a 4 but it is in a different position.
- One number is odd and the other even.
- Both are multiples of three and therefore neither are prime numbers.

I'm sure you can think of other things children might say, and different numbers (–329 and 4·71, let's say) will of course prompt different ideas.

As the teacher you will want to join in and to support the dialogue taking place; this provides a good opportunity to identify terminology and concepts children are secure with, versus things they are struggling to master, and to focus your teaching accordingly. With particular learning objectives in mind you may want to encourage the children to incorporate certain features into their conversations, or to introduce particular things gradually yourself.

Pupil activity 2.2

Order and position
Y1 to Y6

Learning objective: To appreciate the size of numbers relative to other numbers.

Ability to order numbers by size is a key skill and clearly starts with whole numbers and progresses to the inclusion of fractions, decimals, negatives, and so on. In many cases it is useful to include zero.

Having numbers written on mini whiteboards is an ideal mechanism for getting children to order themselves according to the relative sizes of their numbers, since physically ordering the numbers means it is easy to change your mind. Where possible have the children read out the numbers in the order presented as an integral part of checking the order. Numbers can be presented in context (such as length or money) and can be generated in different ways, such as picking a handful of raffle tickets from a pot.

Some ordering activities should ideally also emphasise position, relative to the other numbers involved or to significant points in the number system such as the multiples of 10. It is possible to fix those points first by mutual agreement and for the children to then position their numbers in the gaps roughly where they think they should go. Questioning the children as to why they have positioned them as they have will help to fix the idea that a number is very close to another, about half-way between a couple of numbers, or whatever. When making your selection of numbers, think carefully about what you want to demonstrate.

40 35 39 31 30 34

For example, here we see 30 and 40 and a range in between: 35 to demonstrate the number that sits half-way between them; 34 which is almost but not quite half-way; numbers sitting right next to the multiples of 10 – in other words, numbers designed to generate discussion of such features of the number system. Ability to understand the position of numbers relative to each other underpins rounding.

A washing line and pegs could easily be used as an alternative to the mini white-boards, and clearly all these activities have the potential to be linked to actual differences between numbers.

Pupil activity 2.3

Ladder game
Y1 to Y6

Learning objective: Ability to order numbers whilst developing awareness of the possible numbers in between.

This game builds on children's ability to order numbers, and their developing understanding of the gaps between numbers. A set of number cards (say, from 0 to 99) and a ladder image with several rungs are needed to play this game. Cards should be shuffled and put face down. The rungs need to be far enough apart to fit a number card on each rung and the aim is to fill all the rungs; here we see a ladder with six spaces to be filled. There's a catch, however! The ladder has to have its lowest value at ground level, the next lowest on the first rung, and so on going upwards. So when a number card is picked the children must decide where to position the card (this game draws on probability-related skills) so that each successive card they pick still has a rung it can go on. In the example pictured we were lucky as to the number cards we picked, and we managed to fill all the rungs; we had plenty of games where we weren't so lucky!

This game can of course be adapted, for example, to focus on values 0·1, 0·2, 0·3,…, 9·8, 9·9, 10, or to include negative numbers. Equally it might be a ladder with more rungs, or one with fewer rungs and numbers to 20 only.

Pupil activity 2.4

The hungry crocodile
Y2 to Y4

Learning objective: Ability to determine whether numbers are 'greater than' or 'less than' each other and to use the associated symbols '>' and '<'.

A typical way of introducing the 'greater than' and 'less than' symbols is to use the mouth of a hungry crocodile (or similar) to demonstrate the 'eating' of the larger number, and I have seen many worksheets relating to this theme.

What could often be improved, however, is the choice of numbers. Identifying that $5 < 20$ or that $56 > 7$ is relatively straightforward. Should we want to challenge our pupils' reasoning to a greater extent, it would be better to choose the sorts of numbers that force them to think about their understanding of place value more carefully, hence the numbers I have chosen here. Which symbols would you use?

56 65 21 20 17 70 84 84 51 15

The fourth pair might seem like a bit of a red herring but is there to promote discussion and help focus the children's attention on the use of mathematical symbols, hopefully concluding that in this case we could use the equals sign. Do also encourage your children to articulate the answers they have given in order to relate the symbol to how we would read it.

Eventual work on this theme might include mixtures of whole and decimal numbers, fractions for comparison, or negative numbers.

Pupil activity 2.5

Reinterpreting tens and units
Y1 to Y4

Learning objective: To appreciate that the way in which numbers to 99 are written reflects the numbers of tens and units involved.

The first idea uses foodstuffs as the basis of some tens and units work designed to reinforce the relationships between the size and structure of numbers and how they are written. Whilst my example here uses oranges, it could equally be packets of biscuits or anything else which might feasibly come in multiples of 10.

Pass around an orange and ask the children to imagine the segments inside (depending on experience, you might want to peel one together so that the children know what you're talking about, and you will need some loose segments anyway). How many do they think there could be? All reasonable suggestions accepted, of course, but for the purposes of this lesson, imagine that there are 10. So how many segments would there be if we had two oranges? Three? Using different numbers of oranges and individual segments, illustrate different values, discussing how those numbers would be written down.

This food focus can easily be extended beyond 99. If I have a bag with 10 oranges, or a box with 10 packets of biscuits, I can begin to build larger numbers and compare the physical representation of the number with how it is written. I can also switch my focus to decimals with 2·4 oranges being two whole oranges and almost half (four-tenths) of another.

It need not be foodstuffs used to represent tens and units, of course; other topics can provide inspiration. A topic on 'Bodies' once resulted in numbers represented by different bones!

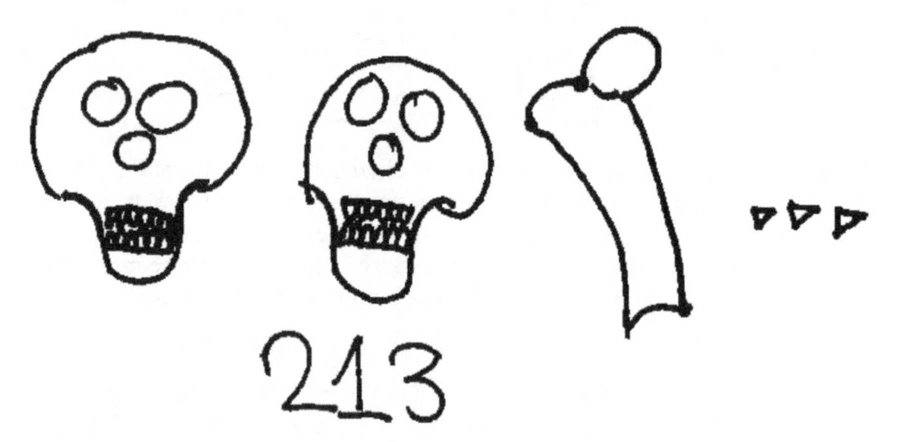

The final idea is a practical one courtesy of a past student. Jaimie suggests combining numbers of stamps (each one hundred), claps (tens) and clicks (units) to represent numbers.

Playing around with the structure of the number system through whatever topic appeals helps to support calculation through offering a more secure understanding of the number system.

Pupil activity 2.6

Digit decisions
Y1 to Y6

Learning objective: To appreciate that the position a digit takes in a number affects the overall value.

A set of digits 0 to 9 (big and bold so that everyone can see them clearly) is great for playing games to develop and fine-tune the children's understanding of place value.

Before each round of the game starts, the type of target number needs to be agreed, for example, that the aim is to make an even three-digit number and to make it as large as you can. One card is selected at random and each player has to decide where they want to put that digit before play proceeds.

H	T	U
8		

Here we see an '8'. This digit would be good as a unit as it is even, but this player has decided to pop it in the hundreds column as the aim is to make the largest possible number – they will be hoping the '9' won't be picked and keeping their fingers crossed that another even number gets drawn! As with activity 2.3, probability skills play a part in deciding where to put each digit based on the likelihood of the other numbers being picked. Target numbers need not be whole numbers and you will want to exemplify different types of number as applicable to the developing understanding of the child.

Pupil activity 2.7

Making and breaking
Y1 to Y3

Learning objective: To understand the value of digits in different positions, with a focus on whole numbers.

Using place value arrows with matching dice is a nice way of practising making two- and three-digit numbers. Having rolled the dice and selected and combined the right arrows, encourage the children to read the number that has been generated and to describe the separate parts that make up the number. Writing this as an addition calculation is a fairly efficient way of recording the mathematics:

$$300 + 50 + 6 = 356$$

Leaving out the tens dice occasionally is a particularly good way of forcing the issue of zero and knowing how to write numbers where one of the columns is empty.

This activity can of course start from the other end, picking a number (raffle tickets are a good source of three-digit numbers) and making it using the arrows in order to help identify its constituent parts.

Pupil activity 2.8

Understanding each digit
Y2 to Y6

Learning objective: To understand the value of digits in different positions, with a focus on decimal numbers.

This activity combines the use of a Gattegno chart and a calculator to focus the children's attention on the different digits that make up a several-digit number. The example here uses a decimal number but could equally focus on whole numbers and could have more or fewer digits. Using the appropriate portion of the Gattegno chart, pairs should select different digits, one of each value. In the case of the example below, the pair would select a unit, a tenth, a hundredth and a thousandth, and discuss the number that the separate parts will generate once combined. Note that chart layout varies; here we see the smallest values in the top row but it could equally be the biggest numbers first.

0·001	0·002	0·003	0·004	0·005	0·006	0·007	0·008	0·009
0·01	0·02	0·03	0·04	0·05	0·06	0·07	0·08	0·09
0·1	0·2	0·3	0·4	0·5	0·6	0·7	0·8	0·9
1	2	3	4	5	6	7	8	9

Once agreed, this number is then typed into the calculator.

9 •	3	7	2

The aim of the game is then to reach zero by taking out one digit at a time according to instructions given by the teacher or another child. This can either be to take out a particular digit if you have one, or to empty a particular column, or a combination of the two. The key to the learning is that the children have to think quite carefully about how to take out a particular digit and replace it with zero: here if I subtract 7 rather than 0·07 it will not help me to reduce my number to zero!

Working in pairs encourages the children to explain to their partner what they plan to subtract to remove a particular digit, but there need to be rules in place to ensure that the children take turns with the calculator.

Pupil activity 2.9

Digit concept cartoon
Y2

Learning objective: To appreciate that digits are reused in different positions and in different combinations to make different numbers.

Concept cartoons can be so easily adapted to meet different learning objectives, and not just in mathematics. This one is designed for Year 2 and provides the children with an opportunity to discuss how a particular number is written with the aim of addressing common misconceptions. Concept cartoons are particularly valuable as children may have the confidence to agree or disagree with a character in the cartoon whereas they may not have the confidence to share their own opinion.

Whilst this example has only one right answer, for many topics it's possible to give more than one correct answer. Variety will avoid assumptions on the children's part that there will only ever be one right answer! It is also possible to vary the number of answers and to provide empty speech bubbles to afford the children the opportunity to devise their own responses.

Pupil activity 2.10

The purpose of the decimal point
Y1 to Y4

Learning objective: Recognition of the purpose of the decimal point.

A nice open-ended task good for generating a need for a decimal point is to have a pile of different coins available to the children and ask them to investigate picking five coins, adding to find the total and recording it before going on to see what other totals they can make with five different coins. Discussion is vital here to cement good practice, but typically occurs naturally as children ask about how to write down combinations of pounds and pence.

Common misconception

'I have £2·59p'

The child here is making a common error in thinking that both units can be in use in the same time. Children should therefore be encouraged to use just one unit at a time, either 259p or ideally £2·59, since this would be more conventional.

Other early experiences could relate to something like how to write down centimetres and millimetres (in centimetres), or metres and centimetres (in metres).

Pupil activity 2.11

Fraction notation and equal parts
Y2 to Y4

Learning objective: To develop understanding of the role of the numerator and denominator in fraction notation, progressing to pairs that total 1. Appreciation of the need for equal parts.

For children to demonstrate their knowledge about fractional notation an open-ended task is ideal. Many schools have a wonderful variety of regular and irregular shapes, and the wider the range to choose from, the better. Children should have the opportunity to select any shape to draw round, to decide for themselves how many equal parts to divide it into and then how many parts to colour. Fraction notation is then used to record the number of parts coloured, and can also focus on the complementary fraction: the number of parts not coloured. This task frees the children from the constraints of a pre-prepared worksheet (noting this is also to your advantage) but offers clear assessment evidence as to their understanding of both fraction notation and the need for equal parts. Remember you can stipulate certain fractions of course, such as insisting that you want at least one example with lots and lots of tiny parts.

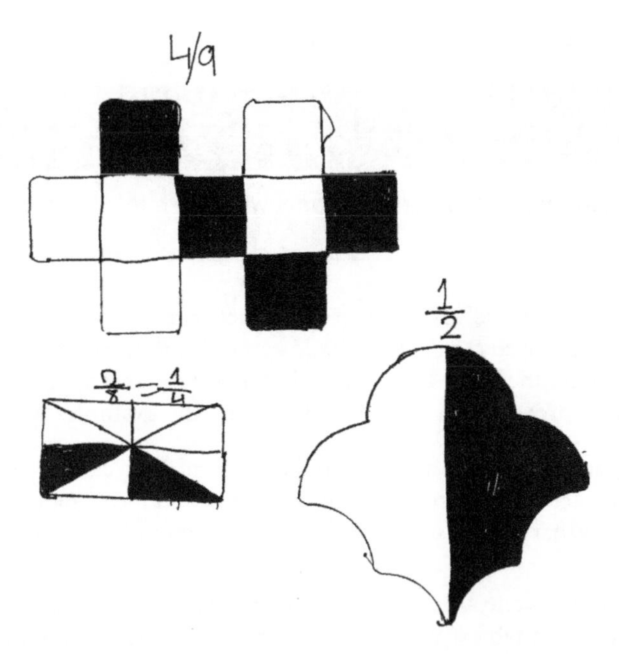

Such an activity can be linked to the food scenarios, where being greedy might relate to eating (or colouring) almost all the parts, or not liking something being illustrated by eating just one part to be polite!

Pupil activity 2.12

Equal portions
Y2 to Y6

Learning objective: To understand that fractions and division are related.

Just as the children above were given the freedom to choose a shape, one might consider a similar approach with regard to fractions of quantity. Counters prove ideal for this activity, with the child picking a handful and counting to find out how many before deciding how to portion the counters (with reminders that the portions have to be the same for fraction work). This has clear links with division work of course, and we might choose to focus on some 'handfuls' like 24 which would work for so many different fractions, or to recognise that numbers like 13 are not going to be ideal for halving or quartering or anything!

This activity can be adapted to focus the children's attention more explicitly on equivalent fractions, with $\frac{12}{24}$ being equivalent to $\frac{1}{2}$ and so on. A context such as a bag of sweets will help to explain this, with eating half a bag of sweets, or a quarter of a bag of sweets, etc. bringing the idea of equivalence to life.

Older children will be expected to find specific fractions of whole-number quantities and also to work with percentages in the same vein. For example, $\frac{5}{8}$ of a bunch of 200 grapes would be 125, whilst 65% of the same bunch of grapes would be 130, just a few more!

Pupil activity 2.13

Fractions problems in context
Y2 to Y6

Learning objective: To develop familiarity with the role of fractions in real-life contexts.

Many contexts lend themselves to illustrating fraction concepts, but the level of difficulty of problems set needs to be considered carefully at the planning stage; the richer the style of problem, the more likely the children are to develop an understanding of how they might approach such problems. Compare the examples presented here:

- In the car park today there are some red cars, some blue cars and some silver cars. Half the cars are blue, there are three silver cars and the remaining quarter of the cars are red. What can we say about the total number of cars and numbers of each colour?
- At a pizza party twenty-four children are going to share eighteen pizzas but can't all sit at the same table. What would be an ideal way of organising the seating so that the pizzas are easy to distribute (and everyone gets an equal share of the pizza)? How much will they get?
- There were ten children in the egg and spoon race on sports day. $\frac{6}{10}$ were girls; how many were boys?

The problems could of course be adapted to make them easier or harder – for example, giving the last fraction in fifths rather than tenths. From time to time I also like to set problems without clear-cut answers. If we did not know how many children took part in the egg and spoon race but we know that $\frac{3}{5}$ of them are girls, then we can consider possible numbers of girls and boys in a more open-ended fashion.

Working collaboratively in small groups and then comparing the different groups' solutions and methods is a good approach for developing flexible ways of working, with problems like the pizza party one ideal because it has several solutions and can be solved in a variety of ways.

What other changes might you make to adapt the problems? What other contexts can you suggest for children of different ages?

Pupil activity 2.14

Equivalent parts
Y3 to Y5

Learning objective: To recognise equivalences within and between fractions and decimals with developing fluency.

Practice makes perfect, as they say, and to revise and retain equivalence facts children will need lots of experience of spotting such relationships. Whereas decimals describe an amount in just one way, with fractions children need to realise that they can involve different numbers and yet still be equivalent. One way of rehearsing these facts is to set up a practical activity where children respond to a question by pointing to one of about four categories displayed on each wall of the classroom, justifying their choice to show how they know it is a valid response.

0·1	0·25	0·5	wholes

For example, a child might point to 0·5 in response to being given the fraction 6/12, explaining how they knew this was equivalent to 0·5. For some children it will help to see the fraction written down to support their making connections between the two forms, and younger children will also benefit from pictorial representations of the decimals and fractions. The categories can be changed and the questions varied depending on the age and ability of your class, with something like 13/52 being harder to relate to 0·25 than 6/12 was to relate to 0·5. I have included whole numbers since children also need to recognise that a fraction like 20/7 involves whole-number parts.

Children could be asked to describe related fractions, for example, that 0·5 as a fraction with 52 as the denominator would be 26/52, going on to make up some of their own relationships, adding cards to the teacher's choices. Including percentages occasionally will also help to prompt discussion of additional learning points, as well as helping the children learn and remember key relationships.

Pupil activity 2.15

Percentages deduced
Y5 to Y6

Learning objective: To appreciate that percentages facts can be logically deduced from known facts.

This activity starts from the premise that if a child knows what 100% is, they ought to be able to work out other percentage facts from this starting point. This does of course rely on secure arithmetic or calculator skills! Numbers can be varied depending on the age and ability of the child, and will ideally draw upon a context as this child has done here.

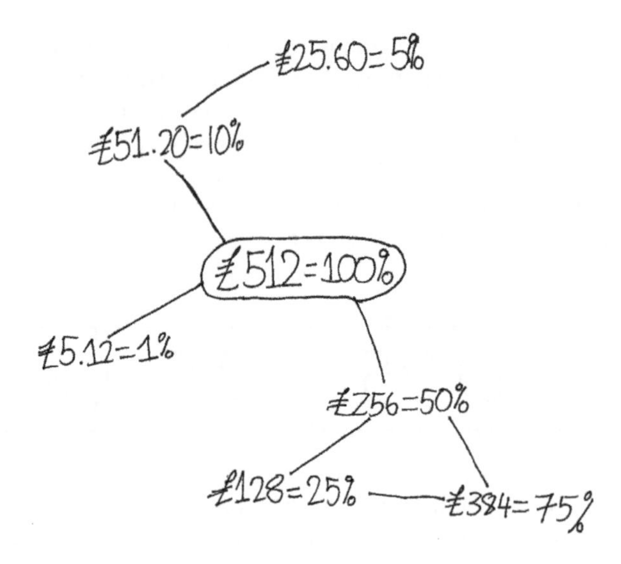

Other examples of possibilities for extending this web include adding 5% and 10% to give 15%; and doubling 10% to get 20%. Alternative contexts to the money one presented here are also possible.

Pupil activity 2.16

Percentage increase and decrease
Y6

Learning objective: To begin to appreciate that percentages are used to describe how numbers increase or decrease, particularly in relation to shopping.

Towards the end of the primary phase, children are often ready to consider the effect of percentage increase or decrease in simple contexts, generally understanding the concept of so many per cent extra free; money off or special offers; or mum and dad bemoaning prices going up! The easiest way to explore this is therefore to present the children with some supermarket type data (real or imagined) and consider what the result of each deal would be...

- Boxes of muesli which are usually 750 g (and cost £2·19) are now 20% extra free. How many grams do I now get for my money?
- Smaller boxes of muesli (500 g) usually cost £1·55 but are currently 20% cheaper. How much do they now cost?
- Luxury muesli is generally more expensive than other brands (£2·09 for 400 g) but I can get three boxes for the price of two. How much muesli will I get for how much money if I am tempted?
- Value muesli is usually a good option but I notice it has suddenly gone up by 30%. It used to cost £0·90 for a 450 g bag. How much does it now cost?

Looking at each scenario as a separate problem will be enough for many children, but the fact that they are related means that there is the potential to compare them to discover the best deal; which would you go for? Is the value product still the best value? Incidentally, like any good problem solver, I may not need to work every problem out to an exact answer to have a pretty shrewd idea that some products are going to be less of a good deal than others.

Further considerations, discussion points suitable for all, help to take the mathematics into the real world. For example, I could ask: Which size box will fit best in my kitchen cupboard? Could I find room for three boxes? Can I afford the biggest box even if it turns out to be the best deal? If I buy the smaller box, will it last until the next time I go shopping? And, equally if not more important, which will taste the nicest?

Pupil activity 2.17

Teaching and learning opportunities
Y1 to Y6

Learning objective: To consolidate understanding of fractions, decimals, ratio and proportion. To develop estimation skills.

Clearly you will have an idea what you want the children to learn, but a simple way of encouraging greater ownership and independence is to give small groups a small pot of coloured sweets (or counters or similar) and ask them how they would use the resource to teach whichever of the topics you wish to them to investigate. Whilst you might set the scene as if the children were to teach younger ones, the act of designing teaching activities will undoubtedly help them to strengthen their own understanding and will afford you some good assessment opportunities in relation to an aspect of fractions, decimals, ratio or proportion.

Estimation activities can also make good use of little pots or clear bags of small items (for example, buttons or conkers) to develop estimation skills. This will initially focus on whole numbers, encouraging sensible guesses followed by tipping out the items and counting employed to check how many. Activities progress to include questions focusing the children's attention on parts: for example, what proportion or fraction of the buttons in the pot do they think are blue; again counting to check, and perhaps contrasting the proportion of blue buttons with the ratio of blues to other colours.

Pupil activity 2.18

Ratio in practice
Y4 to Y6

Learning objective: To recognise ratio in various contexts and begin to understand how to scale up and down.

Paint mixing (either powder paint or ready mixed paint) provides a wonderful opportunity to explore the idea of ratio. Suppose I want to mix green using yellow and blue;

changing the ratio of yellow to blue will clearly result in different shades. Encourage the children to try out some different ratios, recording carefully the numbers of spoonfuls so that they can analyse their results. What does a ratio of 3 parts yellow to 1 part blue look like? What if we swap it over and use more blue?

Another benefit of this scenario is that it is easy to demonstrate the scaling that can be achieved not just by changing the numbers, but by changing the size of the spoon! Clearly if I achieve a green I like and want to paint an A4 picture, a small quantity of paint will be sufficient and I can use the teaspoon, but if I want to paint the whole hall I might be better with bucketloads instead!

Other scenarios which lend themselves to learning about ratio include making squash (where too much water added to a certain amount of juice will result in too watery a taste) and various recipe examples. Another way of exploring scaling through a measures scenario is to tell a story involving a giant: his shoes need to be three times the length of yours; he eats twice as much breakfast cereal as you do; he can walk four times more quickly than you can.

Pupil activity 2.19

Proportions in your class
Y4 to Y6

Learning objective: To understand proportion and the associated language.

The focus here is on proportion related to the children in your class with simple questions about a group such as:

- What proportion of the children have short hair?
- What proportion of the group are boys?
- What proportion of the children are wearing glasses?

Encourage the children to respond using appropriate proportional language, stressing that it is so many children 'out of' the total number. This activity works well with 'true' and 'false' statements since it allows you to model the proportional language you wish to hear; you make a statement and the children have to decide whether, for example, 'three out of the four children are wearing lace-up shoes' is true or false. Once children have grasped the basics, they can suggest their own groups or draw their own pictures and make statements for others to check for accuracy.

Summary of key learning points:

- application to real-life contexts is necessary to ensure learners appreciate purpose – the different ways in which the number system is used;
- ability to move comfortably back and forth in the number system lays important foundations for so much else in mathematics;
- familiarity with the number system has to include part as well as whole numbers;
- sense of quantity is important in relation to how numbers are written;
- it is important to develop understanding of the order and position of numbers relative to each other;
- ability to recognise equivalences, such as between fractions and decimals, is important;
- judicious use of resources can support the development of mental images of the number system.

Self-test

Question 1

The decimal 0·125 is equivalent to (a) 125 hundredths, (b) 125 thousandths, (c) one-eighth

Question 2

The original cost of my house, £71,846, given to three significant figures would be (a) £71,800, (b) £72,000, (c) £71,900

Question 3

If I invest £1000 and it initially grows by 5% but then loses 5%, I end up with (a) the same amount I started with, (b) more than I started with, (c) less than I started with

Question 4

My recipe requires 200 g of flour and 100 g of sugar, but I only have 80 g of sugar, so how much flour should I use? (a) 180 g, (b) 160 g, (c) 150 g

Self-test answers

Q1: (b) and (c) are both correct; the column value gives us 125 thousandths but in its simplest form this is equivalent to the fraction $\frac{1}{8}$, and the easiest way to check this on a calculator is to enter $1 \div 8 = 0{\cdot}125$.

Q2: (a) For significant figures we start on the left and, in this case, take the first three digits as they are, giving £71,800. The final £46 would be rounded down rather than up, being closer to £800 than £900.

Q3: (c) Sadly, this results in having less money at the end, albeit only just less at £997·50. If we reversed the process, losing the 5% first, this actually makes no difference to the final amount as $1000 \times 0{\cdot}95 \times 1{\cdot}05 = 1000 \times 1{\cdot}05 \times 0{\cdot}95$.

Q4: (b) I only have 80 g out of the 100 g of sugar needed; four-fifths of the required amount. Therefore I should only use four-fifths of the flour. 180 g would be considered a common wrong answer with the child thinking that having 20 g less sugar would result in needing 20 g less flour. Remember that ratio involves a multiplicative structure, and therefore addition and subtraction techniques are inappropriate.

Misconceptions

'The numbers go sixty-seven, sixty-eight, sixty-nine, sixty-ten, ...'

'Today is the 21th of June'

'Twenty-one is written 201'

'Fifteen is written as 50 in numbers'

'Sixteen is written as 61 in numbers'

'Before and after the decimal point we have units and tens and hundreds'

'99 is a big number so ninety-ninths must be huge'

'I know eighths are smaller than halves so $\frac{5}{8}$ must be less than $\frac{1}{2}$'

'This decimal number pattern goes 0·5, 0·6, 0·7, 0·8, 0·9, 0·10'

'You can never have more than 100%'

'To make my recipe for more people I need to add the same amount to every ingredient'

'I'm really rich because I've got lots of coins'

'I have £2·59p'

3
Addition and subtraction

About this chapter

This chapter is presented in two different sections:

- Emphasising flexibility in calculation
  - Calculation fluency
  - Mental versus written approaches to calculation
  - Different numbers, different approaches
  - Addition and subtraction structures
- Progression in addition and subtraction
  - Conceptual development and procedural fluency
  - The language and symbols of addition and subtraction
  - The role of counting and number bonds in addition and subtraction
  - Addition and subtraction using number lines
  - Subtraction versus difference
  - Recall and building on known facts
  - Towards formal written methods

The notion of flexibility of approach to calculation is paramount and applies both here and in the chapters on multiplication and division.

What the teacher and children need to know and understand

Emphasising flexibility in calculation

Calculation fluency

However simple or complicated a calculation is, there are three main things that effective mathematicians should aspire to, and that we should want to see our children develop:

- Accuracy – the ability to get the right answer
- Efficiency – the ability to arrive at an answer reasonably quickly
- Flexibility – the ability to change approach according to the numbers involved

It is the last of these that seems to cause the greatest challenge for many practitioners, and yet it is perhaps the most important. Sadly, the difficulties with flexibility can be compounded by curricula with heavy emphasis on particular calculation methods. This is short-sighted since children benefit from being able to pick and choose from different methods as appropriate to the numbers involved in a calculation. This relies on them having an easy relationship with the number system, as explored in Chapter 2, and exposure to different approaches to calculation. This readily happens where children are given plentiful opportunities to share their methods and ideas, ideally in mixed-ability groups, and to find that other mathematicians have different ways of working things out. Developing a repertoire of such skills is also good news from the perspective of accuracy since it allows us to use an alternative approach to check answers we are unsure about.

Lessons where particular methods are introduced and then practised are common; less common, but no less valuable, are opportunities for children to make conscious decisions about which calculations warrant the new method they have learned. Might this be something you could incorporate into your teaching? Might the use of other methods be encouraged for checking answers?

Mental versus written approaches to calculation

Tied in with the notion of flexibility is the need to consider whether a calculation warrants a written approach, or might be tackled perfectly efficiently without resorting to pencil and paper (or a calculator!). Mental approaches are sometimes referred to as the 'first resort', and rightly so. Sadly, I regularly see children completing calculations using onerous written methods: calculations either so simple they already know the answers but are going through the motions as required by the teacher, or to which they *would* know the answers if only they were encouraged to take a closer look. We need to ensure our teaching keeps children's mathematical reasoning skills sharp through the tasks we set and the expectations we have; it is about thinking before acting, as opposed to using standard written methods on automatic pilot. Note that whilst some calculations warrant grabbing that pencil and piece of paper, this is sometimes because it helps to keep track of the numbers involved rather than to carry out a formal written method as such. This is often referred to as 'jottings'.

When thinking about getting children to record their mathematics, we need to consider carefully the purpose of the recording, for example:

- to provide a permanent record;
- to communicate with another mathematician; or
- to support thinking and help work something out.

Whilst the long-term goal for the first two points is that children master the conventions of recorded mathematics, including the use of symbols and standard layout, the same is not true of the third: when all we are after is an answer, personal jottings may be unintelligible to anyone else. Personal jottings should therefore be considered flexible as to layout, and prescription avoided. I am not totally against formal written methods; I just urge you to consider their judicious use and to remember that an estimation of the expected answer will help to ensure you are at least in the right ball park. Sometimes referred to as column addition and subtraction, the vertical approaches are really ideal for working with large numbers and numbers with lots of decimal places, the sort of calculations where it would be hard to hold all the figures in your head as you worked. But equally, if a child chose to use informal jottings, arriving efficiently at an accurate answer, I believe this should be considered an acceptable alternative. Vertical addition is also ideal when calculating the sum of lots of multi-digit numbers; lining them up neatly in their columns is a good idea!

Common misconception

'Addition answers can have two decimal points'

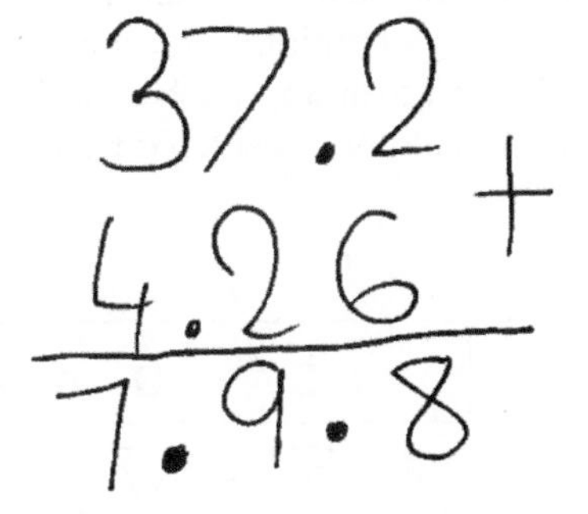

As we can see here, the child has added the two amounts without paying attention to the lining up of the same type of digit; they have also failed to appreciate that the answer (or any number) cannot have two decimal points.

Different numbers, different approaches

To promote flexible approaches to calculation we need to do two main things: to be clear that different approaches are acceptable, and to explicitly draw children's attention to different numbers and encourage them to articulate why they might use different approaches (see, for example, activity 3.1). Note that this variation in approach is predominantly achieved through careful selection of calculations, ensuring there is a good range about which children might be prompted to think differently. This standpoint creates potential tensions where schools limit their teaching to particular methods and expect all children to employ them whatever the calculation.

Let us look at some early subtraction calculations with flexible approaches in mind. How would you anticipate children might tackle the subtraction questions in Figure 3.1? Which calculations might prompt different approaches to others, and why? Such thoughts and discussions form the basis for activity 3.1 and in relation to these calculations we might reasonably expect:

- known answers, such as recognising that 57 – 57 will be zero;
- counting back to subtract small amounts;
- counting up to find the difference between numbers, particularly where the amounts are quite close, as in 57 – 48;
- application of number bonds, for example 57 – 8 = 57 – 7 – 1 = 49;
- a focus just on the units (57 – 6) or the tens (57 – 30), knowing that the other part will not change;
- dealing separately with the tens and units, as in 57 – 12;
- adjustment strategies, such as subtracting 30 rather than 29 and then adding 1 back on;
- application of doubles, for example, I know that double 25 is 50, so 57 – 25 = 32.

You may well be able to think of other ways of tackling these calculations, but remember that accuracy and efficiency need to partner the flexibility I am promoting here. If you find a child trying to complete 57 – 48 by counting back, then they might get the right answer but could easily make a slip, and their approach would not be deemed very efficient for this calculation, whereas it could be considered perfectly acceptable for others.

57 – 29	57 – 25	57 – 6
57 – 8	57 – 30	57 – 57
57 – 12	57 – 0	57 – 48

Figure 3.1 Range of subtraction calculations

The same discussions would be just as profitable were the range of numbers involved different, and it should be noted that larger numbers do not necessarily imply harder calculations. The fact that 2 million plus 3 million is 5 million is potentially more accessible than, say, 46 + 57 and we know that working with really big numbers appeals to children and can give their confidence a real boost.

There are three different laws of arithmetic; these are the 'rules' which govern the ways in which calculations can be performed. They therefore afford us some flexibility as to how we deal with different selections of numbers, particularly when adding. For example, the associative law allows us to take a string of numbers to be added and form associations between different numbers, so for 2 + 7 + 6 + 8 we might spot that 2 + 8 = 10 and use that as our starting point. Equally, we might notice that 2 + 6 = 8 and know that double 8 is 16. What other ways might you advocate? Do your ideas make use of the associative law? We might equally apply the distributive law to addition, which allows us to split numbers and 'distribute' the operation over the separate parts. In the case of adding 57 + 25 we could split the 25 into 20 and 5 and add them separately to the 57. How else could we approach this calculation?

We can use both these laws, though slightly more cautiously, with subtraction. For example, the distributive law allows 57 – 38 to become 57 – 30 – 8, splitting the number to be subtracted (the subtrahend) and calculating in two stages. Whereas formal written methods essentially treat the tens *and* units separately, mental approaches tend to treat the first number (the minuend) holistically.

Presented with a calculation such as 5 + 27 we would hope that children develop an awareness that swapping the numbers to be added (the addends) can make this an easier calculation. This uses the final law of arithmetic, the commutative law. Children will benefit from early experiences to convince them that the order of the numbers in addition calculations does not affect the answer; that 2 + 3 and 3 + 2 do indeed both equal 5. Subtraction is, however, not commutative and less amenable to swapping numbers over: $57 + 3 = 3 + 57$ but $57 - 3 \neq 3 - 57$.

Addition and subtraction structures

We have already begun to consider the possibility of using flexible approaches to calculation looking at the calculations alone, but contexts also give rise to differently structured examples of addition and subtraction, and this can be mirrored in the way the calculations are presented. As teachers, therefore, we have to recognise that structures vary and deliberately expose the children to different examples. If all our questions are essentially of the same type, then children will naturally fail to fully appreciate concepts associated with addition and subtraction.

Studying the problems below, which are all about the lovely flowers my friend Gordon gives me, what differences do you notice? What questions of your own could you add? What contexts would work well with the children you teach? Note the different possibilities for how the scenarios translate into written calculations, and the fact that the unknown is not always at the end:

- I had 5 white and 2 pink carnations – how many altogether? [Two separate quantities combined, 5 + 2 = ?]
- I had a bunch of chrysanthemums. Of the 20 flowers, 13 were orange and the others yellow – how many yellow ones? [Subsets within the original total, 20 – 13 = ? or 13 + ? = 20.]
- Daffodils were growing in Gordon's garden and he gave me a bunch of 20, and then 5 more when they bloomed – how many did I have then? [Original quantity increased, 20 + 5 = ?]
- There were 9 white roses but 2 died so I took them out of the vase – how many were left? [Original number decreased, 9 – 2 = ? or 9 = 2 + ?]
- I have only 3 sweet peas left in the vase; I had to throw 8 on the compost heap as they had wilted – how many did I start with? [Original number to be found, ? – 8 = 3.]
- My small vase had 8 tulips and the big one 15 – how many more tulips did the big vase have? [Comparison of two different quantities, 15 – 8 = ? or 8 + ? = 15.]
- My really big vase holds about 24 blooms but so far I only have 15 – how many more do I need to fill it? [Increase of original amount to reach a target number, 15 + ? = 24.]

Various authors seek to categorise the variety of addition and subtraction problems (see, for example, Barmby *et al.*, 2009) but the most important thing is that we recognise the potential for variety and ensure we incorporate a good range in our teaching to prompt the learner to work flexibly within the different structures. Here we see a good range of examples but none which illustrate more than one unknown; consider the possibility of more open-ended questions with various answers, for example:

- There were 20 flowers in the vase; some were daffodils and some were tulips. How many of each could there have been?

Ensuring you include examples involving zero from time to time is also vital: stories retaining all the flowers or ending up with an empty vase. Setting problems in context helps to give children a rationale for what addition and subtraction are all about and should feature on a regular basis; problem solving is ideally an integral part of a unit of work rather than an add-on at the end. Check that you are using problems in your teaching to give rise to the mathematics and not just as a way of checking that the children can apply the material you have taught them.

Progression in addition and subtraction

Conceptual development and procedural fluency

Clearly children develop mathematically at different rates, and progression in calculation is far from linear; rather, as connections are made, a child's understanding of calculation may be considered more akin to a web of linked ideas. To build these links effectively the emphasis has to be on conceptual understanding of addition and subtraction as opposed to being able to calculate procedurally but without understanding. Extensive prior experience is required if children are to use algorithmic approaches with any depth of understanding, and developing this takes time. Sadly, we know that without time spent laying secure foundations children can make considerable errors when 'blindly' following procedures. Appreciating this underpins the teacher's ability to ensure progression.

Addition and subtraction of fractions is a good example of a facet of calculation which needs to be taught with conceptual understanding in mind. Procedurally, adding $\frac{3}{5}$ and $\frac{4}{7}$ might involve multiplying the 3 by 7 and multiplying the 4 by 5, making the denominator 35, and finally adding the numerators, giving us an answer of $\frac{41}{35}$ or $1\frac{6}{35}$, which may or may not have made complete sense. Perhaps starting with a simpler example, and ideally something set in context, might support children in understanding what we are trying to achieve. If I eat $\frac{3}{5}$ of a scrummy cake one day and $\frac{4}{7}$ of another cake of the same size the next day and want to find out how much I have eaten, the problem becomes a vehicle for exploring the need for common denominators and children can be encouraged to think about suitable numbers and to offer possible solutions. In this case the smallest denominator that will work is 35, with both fractions needing to be changed; in the initial stages I recommend choosing numbers where only one of the fractions needs to be altered to achieve a common denominator. For example, for $\frac{7}{10} - \frac{2}{5}$ the fifths could be changed to the equivalent fraction of $\frac{4}{10}$ and the calculation then becomes manageable. To convince children that the amount of cake has not actually changed, it helps to demonstrate each of the fifths being cut in half to turn two-fifths into four-tenths. Incidentally, do not fall into the trap of not looking closely at the fractions involved and making a calculation more complex than it needs to be. With a little thought $\frac{9}{18} + \frac{4}{8}$ becomes $\frac{1}{2} + \frac{1}{2}$, but not spotting this could result in working with 144 (18 × 8) as your common denominator!

The language and symbols of addition and subtraction

The power of talk should not be underestimated, and where early experiences of addition and subtraction are based in contexts real to the children, the language of addition and subtraction will start to occur naturally. This starts with common words like 'and' to describe the combination of one quantity with another such as '2 green grapes *and* 3 red grapes is 5 grapes'. Clearly the teacher can help to promote mathematical language by modelling additional terms such as 'how many *altogether*' and introducing the concepts of *more* and *less*. Because addition and subtraction can be described using such a variety of terminology (add, plus, minus, subtract, ...) an effective teacher ensures they gradually model a good range, linking new words back to familiar ones. This mirrors reality; the rich heritage of the English language and therefore the benefit of familiarity with the meaning of a range of terms.

'Sum' implies addition, and children's earliest experience of recorded calculation will typically be sums: number sentences describing addition. Children accept fairly readily that '+' is a quick way of recording the fact that two quantities are being combined or a single quantity increased, and soon that '–' relates to decrease. A conscious attempt should also be made to avoid examples which unfortunately obscure what it is you want the children to notice; for example, in the case of $6 - 3 = 3$ the two threes play different roles because of where they appear in the number sentence, roles we want the children to develop an understanding of.

The equals sign is introduced in various ways, but research suggests that rather than 'makes' we might want to stress its role as saying that one side of the number sentence is 'the same as' the other (see, for example, Donaldson *et al.*, 2012).

> **Common misconception**
>
> '$27 + 5 = 27 + 3 = 30 + 2 = 32$'
>
> Be wary of modelling *in*equalities using the equals sign as in the above example where one part of a calculation has run into another and where $27 + 5$ does not equal $27 + 3$ nor does $27 + 3 = 30 + 2$.

Subtraction is often introduced initially as 'taking away', and in terms of conceptual understanding children will benefit from having many real-life contexts which help to illustrate this early sense of reduction: people leave the room or get off the bus so there are fewer people; cars drive away so there are fewer cars; hungry children munch on the cookies so there are fewer cookies. You will no doubt have ideas about contexts which would be familiar to and enjoyed by your children: items magically disappearing or whatever! But, as we saw in 'Addition and subtraction structures', children will also meet other forms of subtraction: for example, if there were 13 children on a bus, 7 of them girls and we wanted to know how many boys, this would not be taking away as such. Aim to gradually incorporate a range of contexts to illustrate subtraction as fully as possible.

In the early stages when modelling the relationship between addition and subtraction of actual items and the recording of this in a number sentence, I would urge slight caution. Whilst the picture in Figure 3.2 could be used to illustrate $3 + 2$ being five strawberries in total, it would be a poor illustration of $3 - 2$ where the emphasis is on taking away, yet it is the type of picture sometimes seen in books and worksheets for this purpose. If, however, the context given is one of difference, then a comparison of the two sets of strawberries becomes more appropriate.

Figure 3.2 Picture of strawberries for modelling addition or difference

Over the years I have sometimes seen practitioners unwittingly laying shaky foundations for later subtraction work by implying that we only ever subtract smaller numbers from larger ones. Clearly this is not the case, but happens as teachers seek to avoid introducing negative numbers too early. In classrooms where children have lots of freedom to explore number, however, for example, through making up their own calculations, the scenario is likely to arise, and in this sort of climate children are generally quite open to the idea of negative numbers. I remember being asked once whether 3 – 5 = 0, which of course it does not, and modelling the taking away of three cubes and wanting to take away two more but not being able to as there weren't enough, hence 3–5 = –2 (read as 'negative' two). Following that conversation, negative numbers started to appear throughout the child's work! Counting back on a number line extending backwards beyond zero would also have been a useful model.

The role of counting and number bonds in addition and subtraction

Early addition and subtraction will typically involve concrete resources and lots of counting, and assumes a secure grasp of cardinality. To become proficient, children gradually progress through a series of steps, initially from counting all to find out how many items there are altogether, for example, when two groups are combined, or finding out how many are left when a certain number of items have been taken away. A slightly greater level of challenge is provided by addition and subtraction involving money. Whilst I might have three coins, how much money I have depends on the values of those coins: lots of different totals from 3p to £6 being possible with current coinage, and more if notes are considered. So calculating with money relies on a firm basis of understanding the value of coins and notes.

Early addition and subtraction based on counting actual items develops to an ability to count on or back from any point in the number system. This is particularly useful for adding small quantities to or subtracting them from a larger number, such as 55 + 3 (56, 57, 58) or 55 – 3 (54, 53, 52), with the last number in the count providing the answer. If you are unsure whether the children in your class are ready to progress from counting all to counting on, a good way of checking this is to count and then hide a number of small items, for example, eight marbles counted into your pocket. Asking how many there will be altogether if you add three more will give you clues as to whether the children are naturally disposed to counting on: thinking about the eight marbles already in the pocket and counting 'nine, ten, eleven' to discover how many there will be in total. If children resort to counting all eight and the additional three, for example, on their fingers, this suggests they are not quite ready for the next stage; though, having said that, experiences of this nature often prompt the skill of counting on. Counting backwards is harder, so it may be some while before children are proficient at this.

Resources such as 100 squares provide models of the number system which can support children's developing understanding of counting on and back, and the 100 square models the idea of adding or subtracting 10 particularly well (see, for example,

Common misconception

'Eight marbles plus three more is ten marbles altogether'

Children giving answers such as 8 + 3 = 10 are making common counting-on errors: 8, 9, 10 instead of 9, 10, 11 in this case. Fortunately, this mistake tends to be easy to spot since answers will always be too small by one: this answer is not 10, but 11. The necessary teaching point is to stress that we already have 8 marbles, so the next number will be 9.

The same misconception can apply to counting back for subtraction, but here the clue will be answers one too large; for example, for 8 – 3, counting back 8, 7, 6 rather than 7, 6, 5.

activity 3.8). An effective teacher will gradually ensure less of an emphasis on counting, in particular in ones, aiming to refine strategies to draw on other aspects of number. With secure number bond knowledge, for example, that 5 is made up of 2 + 3, 4 + 1 and so on, children can begin to recognise that you can jump to significant numbers and then beyond. This might result in a calculation like 27 + 5 being broken down to effectively become 27 + 3 + 2. The technique here is sometimes referred to as 'bridging across a ten'.

Addition and subtraction using number lines

Number lines provide a useful model for step-by-step jumps forwards and backwards in the number system. Early addition and subtraction using number lines will ideally be practical in nature, for example, children physically moving forward and back on numbered carpet tiles laid out in order, and may help to address the counting-on misconception detailed above. Readiness for number line activity depends very much on conceptual understanding of cardinality since addition and subtraction using number lines is considerably more abstract.

Having worked with versions of the number line showing all numbers individually, children will progress towards number lines more continuous in nature and where not all numbers need be shown, such as the 'empty number line' or ENL introduced in Chapter 2. It should be noted that use of the ENL is essentially about informal jottings to aid mental calculation; should a child choose different jumps, or record their workings slightly differently, this should not matter as long as they are getting the right answer in an efficient manner. Note that the addition 'jumps' typically move from left to right whereas subtraction starts on the right (see Figure 3.3), although children do sometimes reverse this.

Where both numbers are larger (or involve decimals), the number line imagery can also be used, typically keeping one of the numbers whole and dealing with the

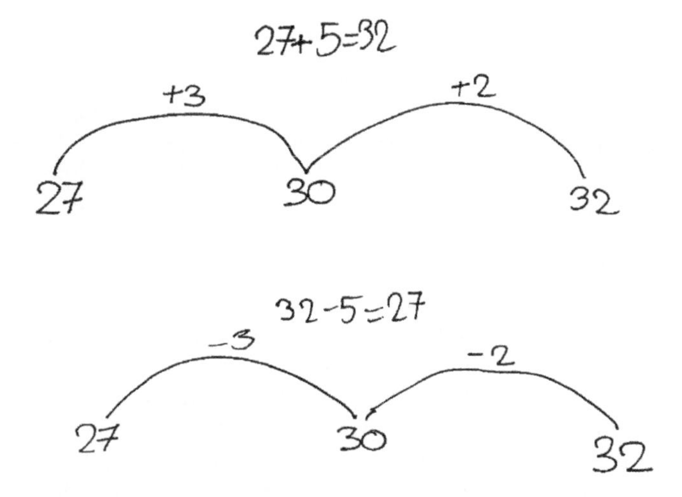

Figure 3.3 ENLs showing 27 + 5 = 32 and 32 – 5 = 27

other in parts. In Figure 3.4 we see counting on from one of the numbers (it could have been the other way round of course) in multiples of hundreds and tens and bridging the final ten in one jump of + 8 rather than having to split this as + 7 + 1.

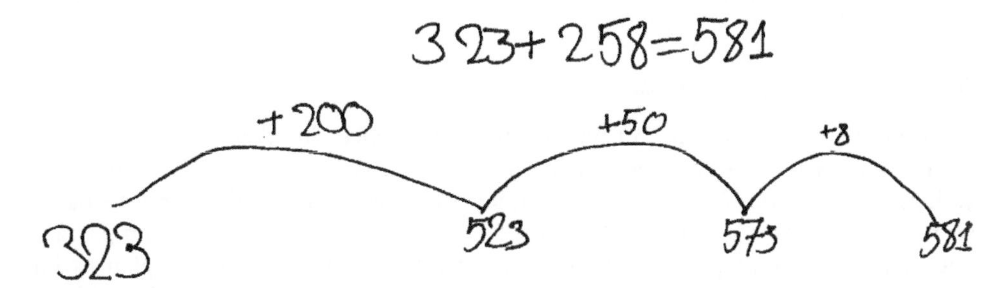

Figure 3.4 323 + 258 shown on an ENL

Thinking about children's journey from use of real or imaginary objects to calculations modelled on ENLs, let us consider a possible stage in between. Both addition and subtraction can be modelled using something like a bead string, giving the number line a more concrete feel. Most bead strings use different colours to denote the tens. The particular advantage for subtraction is that the taking away can be done from either end of the string. In the subtraction we saw in Figure 3.3 the calculation involved moving from right to left, backwards along the ENL, and arriving at the answer 27. Picture this in terms of a line of peas: eating the 32nd pea, the 31st, 30th, 29th and 28th will leave you with 27 peas (you clearly weren't very hungry!). The advantage of working from the right-hand end is that once you have counted back, hopefully in efficient jumps, you have arrived at the answer. In contrast, eating our peas from the other end of the line, the first stage is the easier of the two (I just eat to whatever number I am

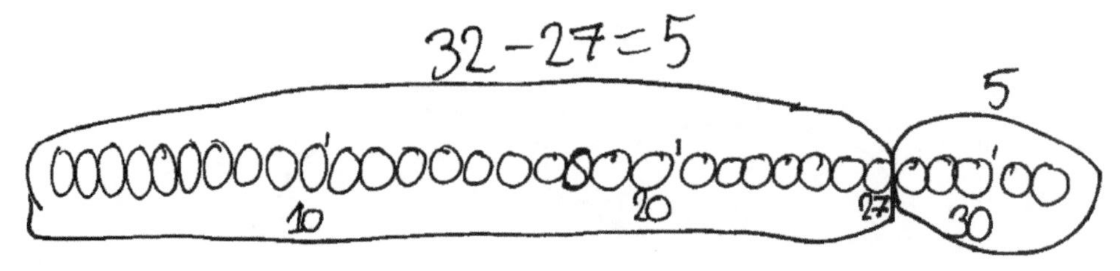

Figure 3.5 Modelling 32 – 27 through the eating of peas!

told to subtract), but to find the answer I need to work out how many peas are left; this is where counting up might be prompted, as seen in Figure 3.5.

Recognising that both are possible is valuable since this affords children flexibility where different numbers are involved; I was clearly very hungry when I ate almost all the peas in Figure 3.5 and it was this that prompted me to avoid the long-winded task of counting back to subtract all 27 peas. For good mathematicians efficiency savings are made through thinking flexibly about calculations. Encouraging children to make choices is therefore important. Thinking back to the discussions about structure, consider the different types of questions one might pose using a food scenario such as this one, including sometimes eating all one's peas, versus someone hating peas and eating none!

Counting forwards to answer a subtraction question is often written about as 'difference' but, as we have seen above, the taking-away aspect can still be made apparent, with a 'how many are left?' feel.

Subtraction versus difference

Children's earliest experience of mathematical difference (in other words, where we are looking for a numerical answer) should ideally be set in a context familiar to the children, and can be enacted or photographed or … as this will make comprehension so much easier (see activity 3.16). Initially difference is likely to be a comparison between two different quantities (towers of different numbers of cubes, or plants with different numbers of leaves or whatever) but can progress to describing single situations and change (for example, the difference between the number of people on the bus at one stop and at another) or differences applied to money (I only have 50p but need 75p to buy my favourite comic; I need 25p more) or to measures (last night the temperature fell to –3°C but today it's much warmer at 14°C, a difference of 17°C).

Common misconception

'The difference between 3 and 11 is that 11 has two numbers'

Essentially this answer is fine since 11 is a two-digit number! And children have been known to give logical answers relating to colour, size, and so on. But asked for the difference in maths lessons we gradually come to appreciate that what we're actually being asked about is the 'gap' between the two numbers: how far apart they are.

Contexts involving food typically work well for comparison and difference; 'he's got more than me' being common for something like numbers of sweets; conversely, 'I've got more than him' for something less desirable (sprouts?!). Whilst mathematical difference helps us to describe how many more or fewer, it should be noted that differences are always positive numbers, whereas subtraction can result in negative answers.

Recall and building on known facts

Being able to remember number facts, such as pairs of numbers or 'bonds', is certainly useful, but recall must never be espoused at the expense of understanding. Being able to pluck an answer out of the air is not enough if the child cannot be totally sure that that answer is right. Mathematics lessons therefore have to include good-quality teaching designed to strengthen understanding, including an ability to work things out, as well as opportunities to practise remembering key facts.

Mathematics is sometimes likened to a huge jigsaw with lots of interconnecting pieces; children who are able to make links between what they are already sure of and new aspects of addition and subtraction certainly benefit from awareness of such connections. This means regularly drawing their attention to such links. For example, the fact that $9 + 4 = 13$ leads me to conclude that $4 + 9 = 13$ as well, using the commutative law. Since addition and subtraction work as inverses to each other, I also know that $13 - 4 = 9$ and $13 - 9 = 4$. Knowing a single fact leads a good mathematician to others, something a colleague refers to as 'buy one, get three free'!

The structure of our number system also makes other connections possible. Instant recall of 10 being $6 + 4$ is a good starting point for the examples below. Note the ways in which the calculations are linked:

- $10 = 6 + 4$
- $100 = 60 + 40$
- $100 = 62 + 38$
- $10 = 6{\cdot}2 + 3{\cdot}8$
- $1000 = 620 + 380$

Common misconception

'Using number bonds $100 = 62 + 48$'

$60 + 40$ does equal 100, but the person in question has overlooked the fact that there will be an extra ten generated by the $2 + 8$. This error could equally well be $10 = 7{\cdot}2 + 3{\cdot}8$ or $1000 = 621 + 489$, for example, and happens when number bonds are secure but used without sufficient thought.

Children often remember doubles more easily than other addition facts, and this can work in our favour in terms of working out other answers: 2 + 2 = 4 and 3 + 3 = 6 and 23 + 23 = 46, for example. You may also have heard some calculations referred to as 'near doubles' where something like 5 + 6 might be based on double 5 plus 1 (or double 6 minus 1).

Another application of known facts is to adjust numbers in ways which make calculations into ones involving known facts. Take, for example, 56 – 29 which becomes 57 – 30 if I add one to both numbers; this is related to maintaining the same difference between the two amounts. Another approach is to subtract 30 from 56 and then to adjust the answer at the end; as I have subtracted one too many, I need to add one back on. Adjustment methods do not suit everyone, however, either because adjusting both numbers does not feel right, or because it is hard to remember which way to adjust the answer at the end. Remember: different methods will suit different people as well as suiting different numbers.

Towards formal written methods

As the numbers involved in calculations get larger, place value begins to play a greater part. Children will benefit from working with place value equipment to appreciate the sizes of the numbers being added and subtracted and the resulting answers, as well as the fact that different calculations focus our attention on the different parts of the number system: 345 – 20 changes the tens digit rather than the hundreds or units, whereas 345 + 101 changes the hundreds and the units columns but leaves the tens alone. Through exploration of physical resources children will begin to realise that you can end up with too many units (or tens, etc.) and that ten of the units could be converted to make another ten, or might not have enough units (or tens, etc.) to complete a subtraction calculation; see activities 3.9 and 3.10 for further ideas.

A growing awareness of these features helps to lay foundations for understanding the relationship between place value and formal written methods. Often referred to as 'algorithms', formal written methods are written procedures ideally used where the numbers are too unwieldy to add or subtract using mental or informal written methods. Such calculations are presented vertically with the column values carefully aligned, and numbers are dealt with one column at a time. Whereas with mental methods we tend to work with the largest value first (and often keep one of the numbers involved whole) written algorithms deal first with the units. Being aware of these subtle differences will help you to ensure you teach them carefully, helping the children to make reasoned comparisons. For more on the topic of calculation, see works by Ian Thompson such as those published in *Mathematics Teaching*, the journal of the Association of Teachers of Mathematics.

Some schools work towards formal algorithms using what they call 'expanded written methods' based on dealing with the parts of the number using a step-by-step approach as we might with the physical resources. I am not, however, a big fan of expecting children to master unnecessarily clumsy recording of these expanded methods. My preference is the emphasis on conceptual understanding supported by resources, developing towards the child making informal jottings as appropriate. This avoids an unnecessary layer of prescription whilst aiming for secure understanding, preceding the introduction of formal written methods.

One such method is vertical addition. The example we see in Figure 3.6 involves two examples of 'carrying' since the units generate an extra ten, and the number of hundreds takes us into the thousands column. Layout can vary slightly, for instance, which side you put the sign (showing in this case that we are adding) or where you record numbers to be 'carried', but this actually does not matter as long as it is clear.

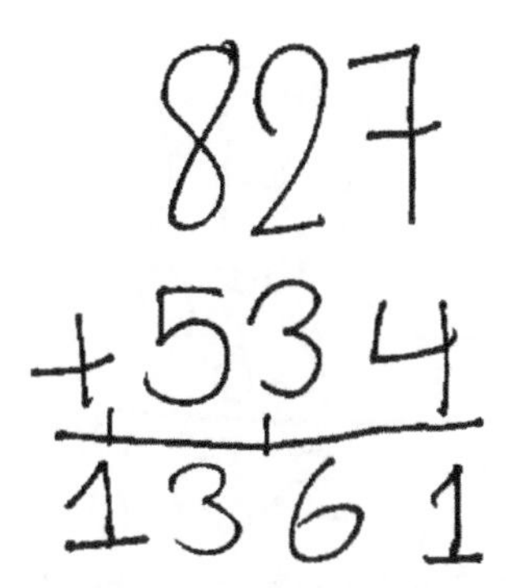

Figure 3.6 Vertical addition of 827 + 534

There are two main methods used for formal written subtraction, and 'decomposition' is shown in Figure 3.7. Here, because we do not have enough units we drop the number of tens from 5 to 4 (50 to 40), exchanging the ten into units. For some time this was commonly called 'borrowing', but the terminology used now tends to be 'exchange' since this better illustrates the process and we do not give back what we have taken. So instead of having 2 units to work with, we now have 12 and can complete the calculation.

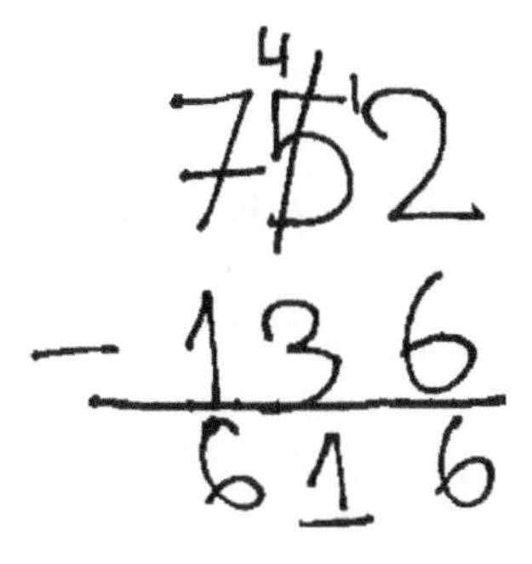

Figure 3.7 Subtraction by decomposition

Using the decomposition approach, it is sometimes necessary to look further to the left in order to begin the exchange process in a column where there is a high enough value – but try to avoid children doing so for a calculation like 1001 – 6, which really ought to be possible to calculate mentally! Try out your own subtraction skills in the self-test section.

A contrasting approach to decomposition is a method referred to as 'equal addition'. In this case if there are not enough units to complete the calculation we just give ourselves an extra ten units so that we do have enough, but having done so need to subtract an extra ten to readjust our answer accordingly. In the example in Figure 3.8 there are not enough tens either so we give ourselves another 10 tens (going from 4 tens to 14 tens) and therefore needed to subtract an extra hundred to balance things out.

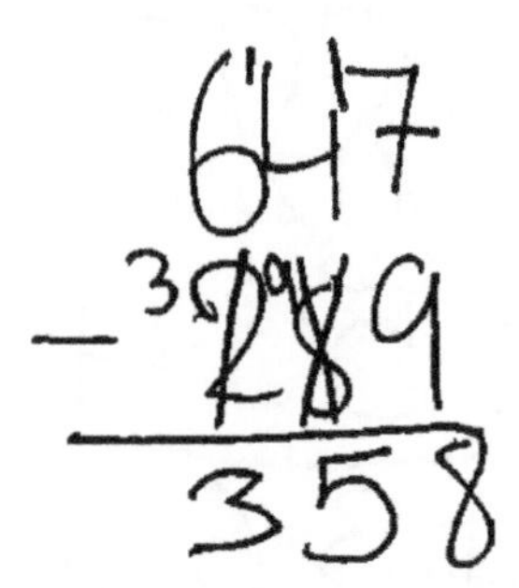

Figure 3.8 Subtraction by equal addition

One of the main issues associated with early use of algorithms is the fact that children can lose sight of the size of the numbers involved since all the digits are essentially treated as if they are units. Mentally I know that three hundred and something plus two hundred and something is going to be in the region of five hundred and something, or possibly over six hundred if there are large numbers of tens involved. I could therefore give a pretty good estimate for 346 + 235 by having a quick look at the numbers involved. Consider, in contrast, the procedure for adding these numbers using the written algorithm described. This can be prone to errors, particularly where 'carrying' is involved.

Common misconceptions

'346 + 235 = 5711 using vertical addition'

3	4	6	
2	3	5	+
5	7	11	

'346 – 218 = 132 using vertical subtraction'

3	4	6	
2	1	8	–
1	3	2	

In the first example the child has added the units correctly but failed to appreciate that we cannot have eleven units, and the overall answer therefore has too many digits and is far too big. The second example exhibits a common error in subtraction: rather than realising that 6 – 8 requires exchange from the tens column to the units (or equal addition), the child has instead swapped the units calculation to 8 – 6 = 2.

Avoiding examples which necessitate carrying in the early stages seems to be a false economy; limiting the range leads to children making false assumptions about the way in which the algorithm works and the transition to the more challenging examples is then problematic.

Activities for addition and subtraction teaching

Many of the following activities can be easily adapted to match different aspects of addition and subtraction or to suit different age groups. Consider which part or parts of your lessons they will prove most useful in, and try out your own ideas.

When using a range of calculations (which aids differentiation), consider presenting them in a random arrangement as we see in activity 3.5. This helps to avoid the pressure on children of feeling they should start with a particular calculation. It can also help to reduce the sense that all calculations must be completed, which can put children off if they feel they will never get to the end.

Pupil activity 3.1

Different numbers, different approaches
Y1 to Y6

Learning objective: To recognise that different calculations may warrant different approaches.

This activity works with children of any age, although clearly the calculations you choose to use will vary according to the ages and abilities of the class. Show a range of carefully selected calculations as seen in Figure 3.1 and give the children a few minutes to think about how they might work out some of the answers. Stress interest in chosen method being greater than arriving at an actual answer. Having shared methods for at least one calculation (aiming to explore at least two different approaches for it), give the children more time to decide how they would work out some of the other calculations, actively encouraging them to think about whether they would work them all out the same way. To start with, individuals may well stick more rigidly to one or two approaches, thus the benefits come from sharing and comparing different children's ideas, really encouraging children to articulate their rationale for their choice of method. We want children to appreciate in the long term that different numbers really do beg different approaches.

As the intention with this activity is for children to focus closely on different ways of working things out when different numbers are involved, I'd recommend focusing on either addition or subtraction rather than trying to do both together.

Pupil activity 3.2

Recall for fluency
Y1 to Y6

Learning objective: To develop fluent recall of basic addition and subtraction facts.

Children who understand the basics of addition and subtraction need opportunities to practise recalling number facts as a regular part of their mathematics lessons in order to benefit from the enhanced fluency. You will no doubt have your own ideas of games children might play, but one of my favourites is called simply '99'. The aim here is to not exceed the target number of 99 when adding cards to a pile and generating a running total. Players are dealt three cards from a regular pack with the jokers removed, and a card from the top of the pack is turned over to initiate play. Players add a card and state the new total, with picture cards (the kings, queens and jacks) counting as tens and the number cards usually worth face value. The only exceptions to this are nines and tens; nines keep the total the same, whereas tens reduce the total by 10. These are therefore useful cards to hang on to until towards the end of the game when you are getting close to the target of 99. Every time it is your turn and you play a card you pick up a new one to replenish your hand; theoretically, if you forget, you play on from that point with only two cards. The game is played over several rounds with players losing a life each time they go over the target. As a family we tend to start with three lives each, and find that our mental skills improve the more we play; discussing different mental strategies employed by different players is interesting and contributes to this improvement.

As you gather other ideas, such as making up your own games using cards or raffle tickets or dice or whatever, ensure you sometimes include zero to make sure that children are not thrown by calculations involving it. Think, too, about opportunities to incorporate fractions and decimals.

Pupil activity 3.3

Doubling up
Y1 to Y6

Learning objective: To develop understanding of doubles and appreciate related facts generated from basic doubles.

To strengthen children's appreciation that doubles can help to work out near-doubles, this activity encourages them to build associated facts from known facts. It assumes they are already reasonably secure with the exact doubles through extensive practical doubling work in the early stages. Before letting the children pick from a range of starting statements, model the sorts of facts which the double might lead to, as in the example here.

700+700=1400
70+70=140
I know that 7+7=14 so I also know...
7+6=13
7+8=15
70+80=150

Although the numbers might vary, this type of activity can be used with all age groups since revisiting doubles periodically helps to ensure that known facts and related facts remain sharp.

Pupil activity 3.4

Addition and subtraction in context
Y1 to Y4

Learning objective: Ability to link calculations with word problems illustrating a variety of structures.

Picking a context and values that would work well for the children in your class, make up a range of problems involving addition and subtraction, plus a range of calculations presented using numbers and symbols. All are presented on separate cards and children should select a word problem and match it to a suitable calculation. An example, with problems designed to link with a literacy topic, is given below. Note that I've prepared a couple of possible calculations for each word problem since there is no fixed way in which they should be interpreted as calculations.

Word problem	Calculation	Calculation
An evil wizard met 32 princes and turned 27 of them into frogs and the rest into newts. How many newts were there?	27 + ? = 32	32–27 = ?
The evil wizard turned 27 princes into frogs on Monday. On Tuesday he turned another 5 princes into frogs. How many frogs were there altogether?	27 + 5 = ?	5 + 27 = ?
There were 32 princes at the ball. The evil wizard decided to turn some of them into frogs but let the others stay as princes.	32 = ? + ?	32 – ? = ?
32 princes were out hunting. 5 of them were caught and turned into frogs. How many princes were lucky and escaped?	32 – 5 = ?	5 + ? = 32
The evil wizard surprised a group of princes. 27 of them were instantly turned into frogs but the wizard left 5 as princes. How many princes were there to start with?	? – 27 = 5	27 + 5 = ?

Rather than varying the values, the numbers have been kept the same to avoid children matching just by recognising the similarity of the numbers involved and to encourage them instead to use their powers of reasoning.

With younger children, however, you might just want to focus on the relationship between the worded problem and the way in which the calculation is presented and might choose to prepare cards which force the children to distinguish between addition and subtraction, for example, choosing between $5 + 3 = ?$ and $5 - 3 = ?$, to match a particular word problem. With older children consider the possibility of incorporating fractions and decimals into the problems.

Ideally children would work in pairs or small groups to prompt discussion, discussion fuelled by the fact that different calculations would be suitable to match some of the word problems or that there are some red herrings. It is for this reason that I would recommend always having more calculation cards than word problems.

Pupil activity 3.5

Calculations as stories
Y2 to Y6

Learning objective: To understand that calculations written using numbers and symbols are a way of representing an addition or subtraction scenario.

Identifying which calculation to undertake based on a word problem is common classroom practice; less common are opportunities for children to work the opposite way. Here we see an addition and a subtraction calculation each turned into a story. The child had free choice as to which calculations to select from quite a diverse mix, noting the inclusion of zero in some of the examples.

56+23=79
43+28=71
25-7=18
75-0=75
100+60=160
3+7=10
562-62=500
15=13+2
51-6=45
36=45-9
98+0=98

Ben went into a shop and bought 13 red sweets then decided to buy 2 more. He had 15 sweets in total.

Billy had 51 marbles. He gave 6 to his friends so he had 45 left.

The examples here are quite simple ones, and the calculations already completed, but the numbers involved can of course be varied to make them more challenging, and the answers omitted. Numerical values can be assigned units to prompt stories relating to measurement.

Once you have a collection of stories generated by the children, these can then be tackled by other children in the class, although of course rather than giving the answers, the children would need to pose questions instead.

Pupil activity 3.6

Estimation and inverses for checking calculations
Y1 to Y6

Learning objective: To develop awareness of reasonable answers.

Whatever strategies a child is using for addition and subtraction, being sure that an answer is right, or at least roughly right, is a useful mathematical skill; both estimation and use of inverses can play a role.

Developing early estimation skills in association with addition can come from simple activities such as dropping a very small number of counters onto a sheet of paper and asking children to estimate how many they think there might be. Sometimes the counters fall naturally into two or more groups; here I saw groups of one and four initially.

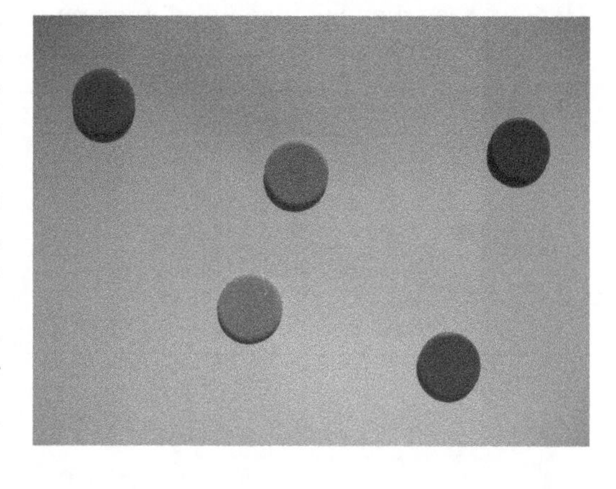

The addition of the smaller groups can then be used as a way of checking how good our estimation skills were, and may involve subitising where we identify how many in each group at a glance rather than counting. Children who develop good subitising and estimation skills are in a stronger position to spot erroneous answers to early addition and subtraction calculations. You might wish to put this to the test by giving a range of completed calculations on separate cards, some right, some wrong, for the children to sort into two piles.

Estimation skills to support more complex addition and subtraction calculations draw in particular on rounding skills (discussed in Chapter 2 in the section on order and position). Knowing roughly the size of the numbers involved by relating them to nearby multiples of 10, 100 or whatever enables a child to perform a simpler addition or subtraction calculation, giving them a ball park figure to check their final answer against. The same sorts of skills can be used when fractions or decimals are involved.

Appreciating that addition and subtraction are inverse operations starts at a young age where practitioners demonstrate practically that adding a certain number of items to get a new total can be undone by then removing those same items (or an equivalent number). This is the basis for using subtraction to check addition calculations or vice versa.

Pupil activity 3.7

Familiarity with the commutative law
Y1 to Y6

Learning objective: To understand the possibility of swapping numbers in addition and appreciate the potential benefits of this.

For children to get to grips with the fact that they can swap the order of the numbers when they are adding, they require lots of experience of doing so, gradually realising and accepting that this does not affect the answer. Explicitly build in opportunities to follow one calculation with the opposite version: for example, having worked out that $3 + 4 = 7$ ask the children to work out $4 + 3$ and then draw the children's attention to the fact that the answers are the same. You will know that they have grasped the principles of the commutative law when, having calculated the first answer, they start answering the opposite version questions without having to calculate them.

The usefulness of this strategy becomes apparent once one of the numbers is larger; adding the smaller of the two amounts by counting on from the larger is much easier than trying to do it the other way round. This can be taught in the context of a puppet who is attempting to add the hard way: can the children help the puppet and make the addition easier?

Once addition strategies other than counting on are developed, the commutative law still has a part to play in potentially making calculations easier, and opportunities should be sought for children to discuss their ideas for the relative ease or difficulty of performing calculations in different orders. Taking a range of calculations, here are some of my ideas:

- To add $563 + 281$ mentally, I would probably still start with the larger of the two numbers and add on the hundreds, tens and units of the smaller number.
- To add $420 + 287$ mentally, I might start from 287; even though 420 is a larger number it has no units so it seems like an easier number to add on.
- When adding a whole number and a decimal, such as $36{\cdot}4 + 20$, I do not think it really matters which way round I do it as I just know that $30 + 20 = 20 + 30 = 50$, but I might try some more whole number and decimal calculations to investigate this further.

Hopefully you are getting the impression that nothing is set in stone. Given different numbers I might well change my mind, but this ability to think flexibility is firmly grounded in understanding of the commutative law.

Pupil activity 3.8

100 square addition and subtraction
Y1 to Y3

Learning objective: To develop strategies for addition and subtraction of tens and units.

36	37
	47
	57
	67

The structure of the 100 square means that children's familiarity with it can aid their ability to add and subtract two-digit numbers either mentally or by tracking the route on the 100 square. Initially the children might develop their familiarity with the structure by completing just snippets of the 100 square, for example, the one here might originally have had only the 36 filled in.

For addition the children's attention should be drawn explicitly to the fact that moves to the right add one, and moving down a row adds a ten, thus: $36 + 1 + 10 + 10 + 10 = 36 + 31 = 67$.

Reversing this process gives us the subtraction equivalent, with moves up and to the left: $67 - 10 - 10 - 10 - 1 = 67 - 31 = 36$.

Working with the full 100 square it is possible to explore the different routes from one number to another, and the pattern of numbers here might well have been 36, 46, 56, 66, 67 instead. Encouraging children to compare their routes and answers with each other will hopefully help to establish that the answers should be the same regardless of whether the tens or units are added first, or even a mixed approach such as 36, 46, 47, 57, 67. Note that calculations such as $31 - 12$ or $38 + 15$ work less well as crossing tens boundaries is not as straightforward, but certainly not insurmountable.

Pupil activity 3.9

Addition and subtraction with place value equipment
Y1 to Y3

Learning objective: To appreciate that the addition and subtraction of units, tens, hundreds and thousands is reflected in the way in which the number is written.

Base 10 equipment is a good way of showing the size of a number in relation to how it is written and is used here as a fun way of rehearsing addition and subtraction involving several-digit numbers. The example here uses three-digit numbers but could equally be just tens and units, or extended to include thousands.

Seat three children in front of the class or group and make the child on the left responsible for the hundreds, the middle child the tens and the right-hand child the units. It is important that they are this way round to reflect the direction in which a three-digit number is written, and if it is possible to seat them so that there is a board behind them on which to write the numbers, even better. The children are then given base 10 apparatus to hold so as to make a number, say, 345, and the composition of the number discussed. Blocks of one, ten or one hundred are then added or taken away and the new number identified each time. Children can, if you wish, work on mini whiteboards to show how they would record the number, and I would also want to check that they could read the numbers properly out loud. In the course of the game there might be certain features which you want to highlight, for example:

- If the tens or units person becomes empty-handed this will be recorded using a zero.
- Rather than presenting the absence of hundreds using a zero, our three-digit number will just become a two-digit number.
- Occasionally we might get a single-digit number (or zero!).
- No one can hold more than nine of their blocks so this may result in increasing the value to the left.
- In contrast, subtraction may involve exchange from left to right.

A sequence like 345, 335, 336, 236, 226, 227, 228, 218, 208, 108, 109, 9, 19, 119, 120, 110, 109 would exemplify all these points.

Once the children get good at this game, it is possible to play slightly faster, and rather than identify the number each time, make several additions or subtractions and see if everyone ends up with the same total at the end. For this it can be especially helpful for the children to have something to write on to keep track of the numbers, and it is possible to add or subtract more than one block at a time.

Pupil activity 3.10

'Exchanges' using place value equipment
Y3 to Y6

Learning objective: To understand the mechanics of formal written algorithms.

Prior to introducing children to formal written algorithms for addition and subtraction, or to support children who are struggling, it is wise for them to have extensive experience of working with place value equipment to support addition and subtraction. The examples we see here were generated by picking raffle tickets and deciding whether to add or subtract, and worked out using base 10 apparatus for support.

AD 213334 446 + AD 213334 451 = 897

AD 213334 302 − AD 213334 190 = 112

AD 213334 155 + AD 213334 487 = 642

Whilst the first one is relatively straightforward, the others require what are typically referred to as ‘exchange’ and ‘carrying’. Rather than shying away from such examples, it is far better to let them occur naturally as part of a rich mathematical diet. This way children realise we sometimes get lots of a particular block when adding, enough, for example, to exchange a heap of units for a ten. Or if there are not enough units to complete a subtraction calculation, children can begin to think about what they would do to solve the problem; this might involve changing a ten into units.

Pupil activity 3.11

Rehearsal of procedure
Y3 to Y6

Learning objective: Increased familiarity with written approaches to calculation.

There are various ways in which children’s exposure to written algorithms can be increased and enhanced. Here we see three possibilities, but you will no doubt have ideas of your own which you can add and the idea can of course be adapted to suit children of different ages and abilities.

- Present the children with several different completed calculations, including some with errors. Ask the children to identify which are right and which are wrong and to articulate how they know. Justifying their decisions to each other will help to test their certainty and check their understanding.

$$\begin{array}{r} 29 \\ 13 + \\ \hline 16 \end{array} \qquad \begin{array}{r} {}^{4}\not{8}{}^{1}4 - \\ 29 \\ \hline 25 \end{array} \qquad \begin{array}{r} 48 \\ 52 + \\ \hline 910 \end{array} \qquad \begin{array}{r} 36 - \\ 28 \\ \hline 12 \end{array} \qquad \begin{array}{r} 27 \\ 13 + \\ \hline 40 \\ {}_{1} \end{array}$$

- Having introduced the children to a formal written method, present the children with a range of calculations and ask them to consider which calculations warrant the use of the new method. As discussed early in the chapter, learning a new method and then applying it is common, but it is far better to give children a range of calculations and for them to actively decide on appropriateness of method; I have purposely included some relatively easy calculations involving reasonably large numbers in the hope that children spot the possibility of working them out mentally. After this we might go onto some which clearly warrant a written approach in order to rehearse the procedure.

$$\begin{array}{r} 139 - \\ 40 \\ \hline \end{array} \qquad \begin{array}{r} 124 - \\ 4 \\ \hline \end{array} \qquad \begin{array}{r} 537 - \\ 37 \\ \hline \end{array} \qquad \begin{array}{r} 461 \\ 86 - \\ \hline \end{array} \qquad \begin{array}{r} 276 \\ 137 - \\ \hline \end{array}$$

- Once children are familiar with at least one vertical method for subtraction, show them calculations recorded in different ways. Can they explain how the other method works, or spot similarities and differences between the numbers?

Pupil activity 3.12

Reasons for column addition
Y4 to Y6

Learning objective: To appreciate why vertical layout is useful.

Till receipts provide us with a ready-made source of lists of figures neatly arranged by column. If the figures were to be added by hand (noting that the till usually does the work) the column arrangement would make our task much easier. Choosing from price tags ranging from 5p to £150, the child has picked five at random and set them out in neat columns to be added:

12p | 8p | £12·99 | £1·05 | £171

```
     12
      8
  12.99
   1.05
 171.00
 ------
£185.24
```

My personal preference would have been to start with the largest amount, perhaps because all the other values would then line up with its digits. The child has done nothing wrong, it is just a possible topic of conversation to gather the views of different children and show that personal preferences are OK in maths.

Children could also look at examples where the numbers have been poorly arranged to help make this point; identifying the pounds and pence in the calculation below is made far more challenging since the numbers are lined up in a rather disorganised fashion.

```
      £ 5 ·9 8
      £ 1 2 ·5 0
  £ 0 ·3 6
          £ 2 ·3 5
          £ 1 0       +
_____________________

_____________________
```

Pupil activity 3.13

Moving back and forth in the number system
Y1 to Y6

Learning objective: Ability to count on and back, initially in ones and then in other amounts.

Early number line activities will ideally be practical in nature, such as a child pretending to be a frog jumping forwards or backwards along a line of numbered lily pads. Rather than recording, the emphasis is likely to be upon discussion relating to numbers started on, how many jumps and in which direction, and finally the number you reach. This early practical work can be developed to include larger jumps, for example, knowing that a jump from 7 to 10 (or vice versa) is a jump of 3; jumps of 10 can of course be related to the addition and subtraction activities using the 100 square as seen in activity 3.8. As a contrast to jumps, you might at other times want to focus the children's attention on removing items from one end of a number line or the other as illustrated in the earlier discussion about eating peas! (See the section on addition and subtraction using number lines.)

With older children, once decimals are included in addition and subtraction, number lines continue to prove a valuable tool for keeping track of jumps forwards and backwards in the number system. The examples here, 14·5 – 3·8 and 23·6 + 10·4, were generated using a couple of spinners and solved using a number line. Numbers involved could of course include more decimal places, or be simplified to have decimals on only one spinner.

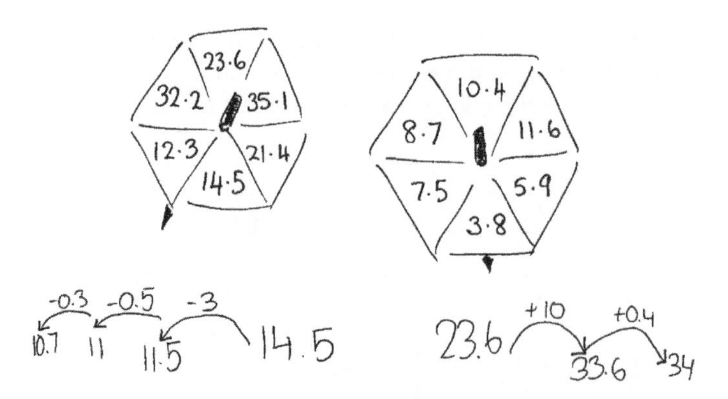

Number lines also lend themselves to working with negative numbers where the same sorts of questions can be asked relating to starting number, number of jumps and so on. There might well be a context here, such as temperature, and the number lines could be presented vertically rather than horizontally.

Pupil activity 3.14

Different ways of paying
Y1 to Y5

Learning objective: To understand that coins and notes can be combined to pay amounts in different ways.

Working out possible ways of paying for an item costing a particular amount lends itself to an investigative approach, and whilst the addition itself may be straightforward for the older children, the opportunity to do things like working systematically is still beneficial. I just checked the cost of bananas on my supermarket receipt. One cost about 10p and had I bought just one on its own, I could have paid for it in a number of different ways, eleven in total, from all pennies to a 10p coin with combinations of 1p, 2p and 5p in between.

Here we see a child who has worked out that there are five different ways to pay for a 6p carrot, but who might be encouraged to find and record the possibilities more systematically, perhaps by listening to other children or the teacher articulate how they knew they had found all possible combinations.

6p

5p + 1p
1p + 1p + 2p + 2p
2p + 2p + 2p
1p + 1p + 1p + 1p + 1p + 1p
1p + 1p + 1p + 1p + 2p

As costs increase, detailing all solutions is less practical and children might instead be asked to suggest the three best ways to pay. For example, 21p might be 20p + 1p, or 10p + 10p + 1p, or 10p + 5p + 5p + 1p. Another investigative option is to look for amounts which can be paid using a certain number of coins. We have just seen that 21p can be paid using three coins (10p + 10p + 1p). Which other amounts can also be paid using exactly three coins? Are there any amounts for which three coins will not work? How do you know?

Whilst the focus here has been on adding to make exact amounts to be paid, activity 3.15 takes the next logical step, and many children may make this leap unprompted.

Pupil activity 3.15

Calculating change
Y2 to Y5

Learning objective: Ability to suggest how to pay without the exact money and to calculate the change.

Once children have got to grips with the fact that different coins have different values, and are able to add such values in order to pay for items using the exact money, attention often turns to working out how much change you would get if you handed over too much money. I find it helps if you model not having the right coins to pay an amount exactly, encouraging the children to offer a solution to your problem. This helps them to understand the whole concept of change.

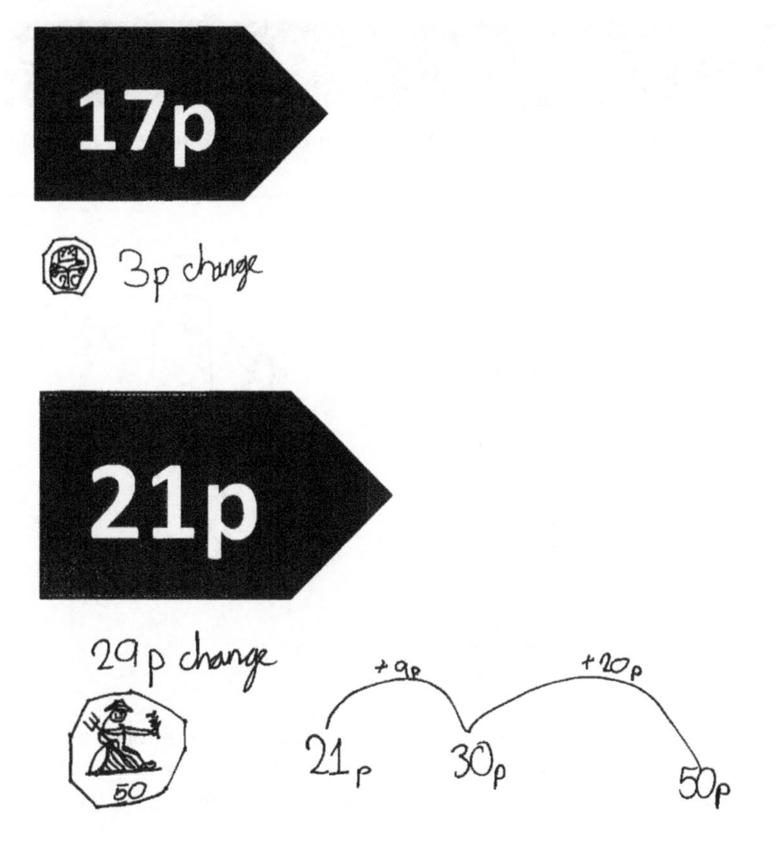

The key feature with this activity is that rather than being told how to pay for the amount, the children themselves have to decide which single coin they could use to pay, stressing the fact that this needs to be a coin (or note) worth more than the amount owed! A number line model is ideal for illustrating the fact that the shop keeper wants to take money up to the amount you owe, counting on to find out how much you are due to get back. I find this approach preferable to using more of a subtraction approach, but it does depend on the numbers involved. My first choice of 17p was deliberate as most children will recognise 17 as being quite close to 20, whereas 21 provides more of a challenge as it is a long way from the next suitable coin, though of course you could use more than one, such as 20p and 10p. From an organisational perspective, having introduced the task carefully and gauged the children's grasp of the concepts involved, I would ideally give the children a big choice of price labels; those who need to secure their understanding can pick similar or even simpler examples, such as 7p or 18p, whereas others might opt for 57p, 99p, £1.90. This way differentiation is inbuilt in the choices the children are able to make.

Whilst older children should be comfortable with the concept of change, I find they sometimes still need reminders and support with approaches to its calculation. You may therefore wish to incorporate elements of calculation of change into multi-step problems from time to time.

Pupil activity 3.16

Difference
Y1 to Y6

Learning objective: To understand the concept of difference and its application to different scenarios.

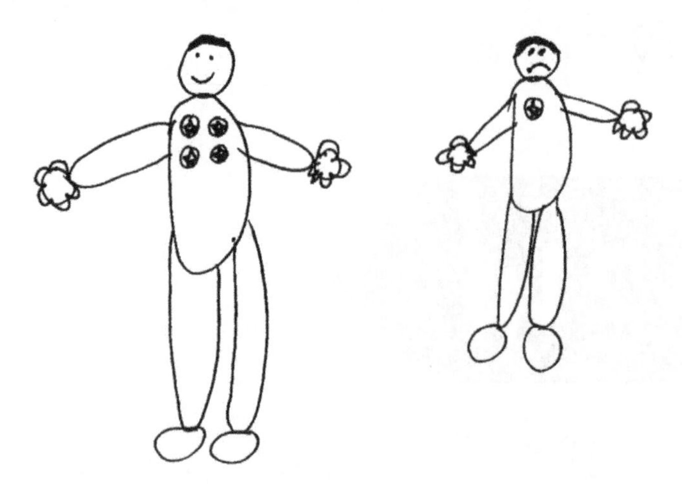

Although here we see a picture of children with different numbers of stickers, early opportunities to compare two different amounts will ideally be practical: real children with real stickers, or anything else that can be compared numerically. The initial focus, however, will just be on comparison rather than working out the difference as it is important to establish key concepts relating to whether children have the same or different numbers of stickers. Encourage the children to articulate what they notice but expect variety in the language here and praise the contributions whilst modelling accurate alternatives yourself, such as talking about children with more or fewer stickers than each other. The final step towards younger children appreciating numerical difference is to begin to ask 'How many more?' when comparing collections of real items, such as two bowls with different numbers of pieces of fruit. Personally I'd start by modelling two types of example:

- big differences like a bowl with a couple of apples versus another with lots (because a difference in number will be obvious even if we do not know what the difference is);
- differences of only one to encourage statements about one more or one fewer (perhaps the easiest numerical difference to talk about).

To help children appreciate the differences it helps to line up the items (or take the fruit out of the bowl or whatever) so that the 'sameness' can be easily spotted and the extras can be more easily counted. Once children understand the basic concept of difference, numbers can of course become gradually more challenging and children can be encouraged to rely more on their calculation skills and less on counting the difference between real numbers of items. This need not be interpreted only as subtraction since many children will find it easier to count up to find the difference between two quantities rather than subtracting *per se*.

To identify differences between fractions, the fractions ideally need to have common denominators or be converted to fractions with a common denominator. Try to introduce such work through problems in real-life contexts, such as wanting to identify how much more pizza one friend has eaten than another. If someone has eaten seven-tenths and another person three-fifths and we want to know who has eaten the most, this may spark a reason for working with common denominators and lead children towards being able to calculate the difference.

As we saw in activity 3.13, number lines are great for supporting children in moving around the number system; with difference they provide a helpful visual image for seeing how big the gap is between the two numbers you wish to compare, and this is particularly helpful for differences in temperature which span positive and negative values.

Difference can be exemplified in many different ways, but try over the long term to work from measurement scenarios as well as purely number ones. Here the word 'less' comes into its own; were we to want to compare two quantities of (say) custard, we could talk about one of the jugs holding more custard, whilst the other jug holds less. The difference in this scenario would therefore be given in something like millilitres. 'Fewer' relates to countable items, so if we were referring to something like rice we could talk about fewer grains, but in general would discount the possibility of counting and just refer to one person wanting less rice than another, and the difference could be described in terms of mass. Where contexts relate to measures (for example, comparing the height of sunflowers grown by the children, or charting the growth of a single sunflower from week to week and working out the difference in height), children will therefore need to appreciate the units involved. Older children might pursue more complex investigations such as comparing the carbohydrate intake of different athletes.

Pupil activity 3.17

Advanced concepts of difference
Y3 to Y6

Learning objective: To understand that concepts of difference can also be applied even when unknowns are involved.

This activity is based on the idea of comparing two children's pocket money without actually knowing how much they started with. Read the story outlined below. Who ends up with the most money? How do you know?

- Toby and Rebecca each have the same amount of pocket money in their piggy banks to start with.
- Grandma visits and gives both children £2 to add to their savings.
- When they are out the next day Rebecca decides to buy her favourite magazine. It costs £1·99.
- Toby doesn't buy anything. He's saving for a new computer game which costs £18 and he doesn't yet have enough money.
- Dad wants the car washed when they get home. Rebecca is not interested as she wants to read her new magazine. Toby offers and he does such a thorough job that his Dad pays him £5 and he now has enough money to order his new computer game.

You may well have some thoughts about the children's starting amount and if you want to test out your ideas try question 4 in the self-test section. It is Toby's story that contributes the clues as to the limits (how much or how little money he could have started with), but we will never know how much the starting amount actually was and some children will struggle with this idea. Working through from different amounts is a way forward if children are stuck: finding that some starting amounts would not be enough; that there are different amounts that will work; and, most subtle of all, that Toby must have started with less than a certain amount, otherwise he would have been able to afford his new game sooner. What we can be certain of is that Rebecca has more money than her brother at the end of the story and more money than she started with, albeit only 1p! Many learners find an ENL image useful in charting the increases and decreases in money, as we see here for Rebecca's story.

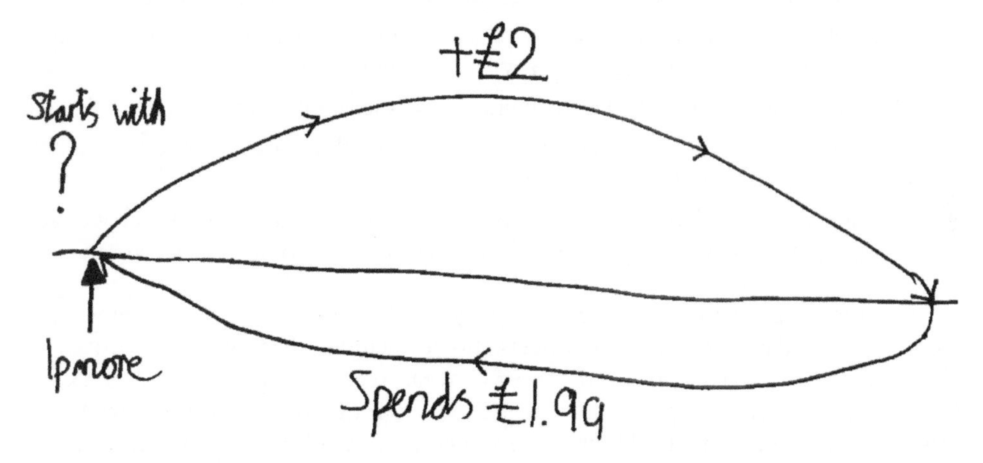

This would be a particularly complicated story to start with and could be simplified by the children not dipping into their original savings (both ending up with more than they started with) or by avoiding a scenario involving units, in this case money. This style of activity should be considered a group task; children will need plenty of time to think the story through and talk with each other about their ideas, and I feel access to something to write on is essential. Any scenario where amounts go up and down will lend itself to starting with an unknown and provide material for exploring this more advanced concept of difference, which links to Chapter 6 and elements of algebraic thinking.

Pupil activity 3.18

Thoughtful calculation
Y1 to Y6

Learning objective: Ability to approach calculation with flexibility in mind.

Pages of 'sums' have their place in mathematics, but with a little thought can be used far more effectively. If your children are in the habit of working through systematically from the start and using the same method to answer each question, then think again! Consider the following suggestions as possibilities:

- Get the children to scan the sheet before they put pencil to paper. Are there any calculations they think they already know the answers to? How sure are they? Can they convince a friend that their answer is right?
- Of the other calculations, which do they think is going to be the hardest and why? What sort of method do they think they might try to solve it?
- Which of the questions do they think are going to have particularly small answers? Pick a couple of them and work them out to check which is the smallest.
- What about large answers? Do any of the calculations take us from tens to hundreds or hundreds to thousands? Try one of them to see.
- Having worked a few out, did the children use the same method each time?
- Can the children spot any calculations which they think are going to have a zero in the units/tens/hundreds column?
- Can the children identify a calculation that is going to have an odd answer and another which is going to be even? Were they right and how did they know?

Approaching calculations in the ways suggested here is particularly advantageous from a differentiation perspective. Sheets should include a wide variety of questions for the children to choose from; the most able can seek out calculations which challenge them without having to first work their way through other questions, whilst the least able can select manageable questions. The very act of choosing which question to answer has considerable learning potential, since the learner must look more closely at the range in order to decide.

Any page of calculations lends itself to this type of approach, and can include addition and subtraction, whole numbers and decimals, and calculations presented both horizontally and vertically.

Summary of key learning points:

- as calculations vary, different approaches to calculation are warranted and desirable;
- appreciating that addition and subtraction structures vary should ensure use of a rich range in our teaching;
- real-life contexts will support children in making sense of addition and subtraction;
- procedural fluency requires secure counting skills and ease with basic number bonds and place value;
- recording of calculations should be purposeful and may involve personal jottings.

Self-test

Question 1

The same calculation has been done using two different methods: decomposition (shown on the left) and equal addition (on the right) but since they show different answers they cannot both be right.

```
 [2]
  3 [1]2 [1]2 . [1]0          3 [1]2 [1]2 . [1]0
        6    4 .   7       -  1 [7]6 [5]4  .   7
-  ______________          ___________________
     2  6  8 . 3               2  5  7  .  3
```

I think it is the decomposition that has gone wrong. True or false?

Question 2

8 – 3 + 2 – 5 = 8 + 2 – 8 = 2. True or false?

Question 3

8 – (3 + 2) – 5 = 8 – 3 + 2 – 5 = 2. True or false?

Question 4

In activity 3.17 we heard Toby's story. A child who had some savings (but we do not know exactly how much money), was given £2 by Grandma, did not yet have enough money to buy a new computer game (£18), but then earned £5 washing the car for Dad and could then afford the game. Toby could have started with anything from £11 to £16. True or false?

Self-test answers

Q1: (True) For this decomposition example to work, exchange is required all the way across, thus both the 12s should actually only be 11s.

Q2: (True) The use of the associative law here is fine as the values to be added and subtracted have been retained with – 3 and – 5 resulting in the subtraction of 8.

Q3: (False) The difference here is that the brackets direct us to work out what's inside them first. This gives us 8 – 5 – 5 = –2.

Q4: (False) It is certainly true that Toby had to have at least £11, anything less and the £7 he got from Grandma and Dad would not have been enough to allow him to purchase the game. To identify the falseness of the statement we need to consider the £16; had Toby had £16 to start with, as soon as Grandma gave him £2 he would have had enough money to buy the game. He must therefore have started with less than £16, even if it was only a penny less! Therefore the limits for the unknown starting amount must be £11 and £15·99.

Misconceptions

'Addition answers can have two decimal points'

'27 + 5 = 27 + 3 = 30 + 2 = 32'

'Eight marbles plus three more is ten marbles altogether'

'The difference between 3 and 11 is that 11 has two numbers'

'Using number bonds 100 = 62 + 48'

'346 + 235 = 5711 using vertical addition'

'346 – 218 = 132 using vertical subtraction'

4
Multiplication

About this chapter

This chapter describes the essential knowledge underpinning effective teaching of multiplication. It is organised in two main sections, subdivided as follows:

- Towards conceptual understanding of multiplication
  - Origins of multiplication in real life
  - The influences of the laws of arithmetic
  - Imagery to support understanding: arrays
  - Fundamental multiplication facts
- The language, resources and methods of multiplication
  - Language and symbols
  - Number system resources
  - Working with fractions and decimals
  - Place value resources
  - Written approaches to multiplication

What the teacher and children need to know and understand

Towards conceptual understanding of multiplication

Origins of multiplication in real life

Multiplication is sometimes referred to as 'repeated addition', and this idea of repetition is fairly central to multiplication, particularly in the early stages, and is what we predominantly concentrate on here. It is about having a group or set of a fixed size (picture some bags, each containing six apples) and thinking about how many altogether once you know how many of the group you have got. So if I pick up three of these bags of apples in the supermarket I will have 18 apples in all.

A second facet of multiplication relates to the concept of scaling; an idea we met through ratio in Chapter 2 and explored in the context of adapting recipes. Ingredients could be increased or decreased by different scale factors, with multiplication the key operator.

Whether we are concerned with scaling or repeated addition, the use of context in teaching is vital – contexts which offer up multiplicative potential, something which Multilink can never do. If I ask you to imagine a tricycle, something familiar to most young children, and to think about how many wheels it has, you can hopefully picture one with its three wheels. What if there were two tricycles, can you picture those? How many wheels now?

Common misconception

'I have used multilink to work it out and I think $3 \times 2 = 5$'

With early experiences of addition embedded, children's early answers to multiplication calculations often suggest that they have added rather than multiplied. This happens far less if the problems are set in context as it is harder to give an answer of 5 if you are imagining two tricycles, each with three wheels.

The tricycle scenario can of course be modelled with actual tricycles, or at least pictures of them to support children's visual images and help them to appreciate the concept of multiplication; that it is about replicating groups of the same size. In Figure 4.1 the child has chosen an aeroplane to illustrate multiples of 3 and drawn other vehicles depicting other numbers of wheels to support his multiplication work.

Many topics lend themselves to generating a range of multiples, and minibeasts are a particularly rich topic, one we explore in activity 4.1. You will no doubt be able to think of other ideas, such as flowers with different numbers of petals; so many biscuits per packet; body parts (particularly good for twos) and even some mathematics equipment such as

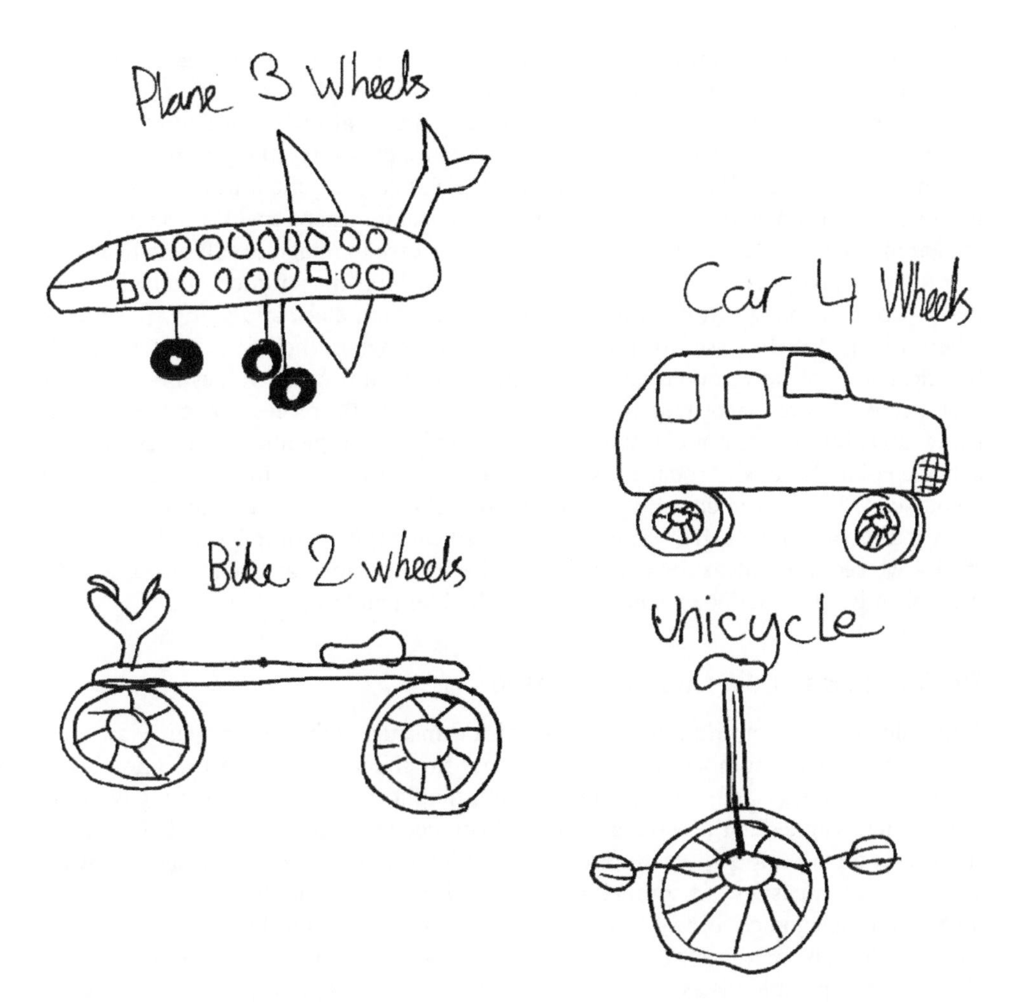

Figure 4.1 Vehicles with different numbers of wheels lend themselves to teaching multiplication

Numicon where if I have lots of yellow pieces I am working with multiples of 3. Shapes work in just the same way; a heap of pentagons will illustrate the five times table if we are thinking about numbers of sides or corners. Whatever your own ideas, check each time that the item is inherently multiplicative: that it has a distinct grouping of a certain size.

To return to my original suggestion that resources such as Multilink cubes are far from ideal for modelling early multiplication, this is precisely because they do not have any inbuilt groups of a certain size. Rather than being predetermined, such groups have to be generated by joining the cubes together, and this is where children can be very confused about how to interpret a calculation: does 3×2 mean I should make twos or threes with the Multilink? How many of each? Or should I just make a two and a three? It should really be no surprise that we get wrong answers in early multiplication if we are using resources with no inherent suitability. Things like cubes therefore have more potential at a later stage, once children have a developed a more secure conceptual understanding of what multiplication is all about.

Illustrating multiplication through real-life contexts affords practitioners opportunities to highlight particular cases which can cause children difficulty, notably multiplication by 0 and 1. Thinking back to the vehicles context, the unicycle would help to illustrate multiples of 1, but we might decide, quite intentionally, to incorporate something like a sledge; however many sledges we care to imagine, still no wheels!

All the resources suggested so far have something in common: that whatever makes up the groups can be counted – the wheels, the petals, and so on. Resources where this is not the case are therefore more challenging, including money, which has clear multiplicative potential but children have to appreciate values first: the fact that a coin or note is worth something. Thus a pile of 10p pieces can model the ten times table, but we cannot count each of the individual ten pennies as represented by a single coin. Cuisenaire rods, a mathematics resource available in many schools, are colour-coded like Numicon, but as with money children first have to appreciate that each rod has a value, as the number represented cannot be counted as such. In activity 4.4 we see some ideas for using junk packaging, a readily available resource with multiplicative potential, best suited to use with older children.

The influences of the laws of arithmetic

Multiplication is a commutative process; this means, using the commutative law, that the order in which numbers are multiplied has no effect on the answer. Thus $3 \times 2 = 2 \times 3 = 6$, and the same principle applies to multiplication with however many numbers, which is linked to the associative law. If you need to convince yourself you might want to compare $2 \times 3 \times 4$ with $2 \times 4 \times 3$ and $3 \times 4 \times 2$ and so on – fingers crossed your answer equals 24 every time! The associative law is rather like numbers having a change of allegiance; use of brackets in mathematics helps to indicate calculations to be dealt with first, but where only multiplication is involved, it does not matter. Because multiplication is associative $(2 \times 3) \times 4 = 2 \times (3 \times 4)$ and so on; the 3 is associated with the 2 to the left of the equals sign but with the 4 on the right and either is fine as we get the same answer. For more on this, see activity 4.17.

The commutative law suggests that order does not matter, and in many ways it does not, but as a teacher of mathematics you might be interested to know that 2×3 can be interpreted differently than 3×2 mathematically speaking, with the pictures in Figure 4.2 illustrating the difference.

The reasoning behind the order is that the first number tells us the group size, and the second tells us the number of times the group appears. Thus the picture on the left is three balloons twice, whereas the picture on the right shows the balloons grouped in twos and there are three groups altogether. If the order goes against what you have always believed then you are in good company, and I believe the lack of awareness stems from the interpretation of '×' as 'lots of' rather than 'times': the number of times the group appears. Clearly, if you want to draw children's attention to the possible interpretations of multiplication in relation to different scenarios or pictures like the ones in Figure 4.2, then you need to avoid examples such as $2 \times 2 = 4$ which would obscure this.

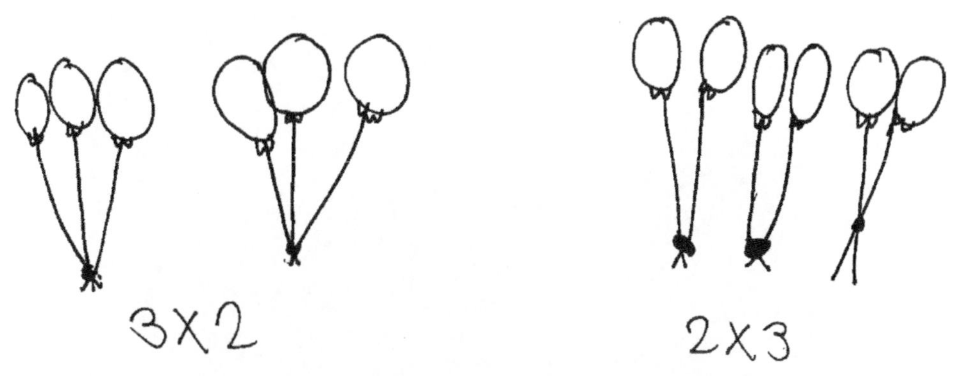

Figure 4.2 3 × 2 and 2 × 3 illustrated pictorially by bunches of balloons

Given that however we interpret 2 × 3 (or 3 × 2) it will still equal 6, I have never let the distinction bother me too much in my teaching and am happy to accept either interpretation. Some years ago we discovered that one of my nephew's teachers was not quite as enlightened. Zachary's homework was deemed incorrect despite being mathematically accurate, and it was because his pictorial representation of the calculations didn't reflect 'lots of'.

The distributive law is also useful when we are multiplying and is what many mental methods, plus most written methods, are based on. Take the calculation 24 × 13 and consider how you might work it out mentally. If you decided to think about 24 × 10 (240) and 24 × 3 (72) and then added the answers together (312) then you used the distributive law to split the 13 into 10 + 3 before multiplying each part separately. You could just as easily have split the 24, effectively making the calculation (20 × 13) + (4 × 13) = 312. The split need not be a separation of tens from units, although this does tend to be the case with both formal and informal written methods. Consider why (12 × 13) + (12 × 13) would also work, and why somebody might choose to use the distributive law in this way. Activities 4.11 and 4.18 both make explicit use of the distributive law.

Imagery to support understanding: arrays

The term 'array' refers to a rectangular arrangement (of dots or squares or whatever) where the rows and columns represent the numbers involved in a multiplication calculation. Here, too, we might draw on familiar, everyday imagery such as chocolate bars, and in Figure 4.3 we see a bar with 24 chunks, with 8 in each row and 3 in each column.

Arrays are ideal for demonstrating the commutative law since whether we focus on the rows or the columns, the answer will still be the same. Whereas in the early stages we might seek to avoid examples where numbers are repeated, arrays model square numbers rather nicely since they can actually be shown to be square! Activity 4.16 gives children the opportunity to hunt for square numbers as part of their exploration of arrays.

Figure 4.3 Chocolate bar illustrating $8 \times 3 = 3 \times 8 = 24$

Once children's understanding of multiplication is secure, they can begin to apply it to topics such as area and also algebra. Early work on area tends to focus on rectangles containing the rows and columns of an array; perhaps unit squares 1cm^2 each. At this stage, because the unit squares are visible the area can be found by counting the number of squares, but you will want to explicitly draw the children's attention to the multiplication aspect: that rather than counting every square I only need count how many are in a row, and how many in a column, and can then multiply to give the overall total. Using a 'length × width' formula builds on this, focusing on rectangles where there are no unit squares inscribed; this forces us to use multiplication as we cannot resort to counting the squares. Having said that, we might want to show that we could put in the centimetre markers along each side and join the lines, creating the internal array and proving the match between the number of squares and the answer obtained using the formula. But the long-term aim is that the children can understand and use the formula, gradually extending its use to find the area of rectilinear shapes and triangles. Algebra helps us to deal with unknowns and variables (for more on this, see Chapter 6) and the area model helps here; if we know that one edge of the rectangle measures $4a + 3$ and the other $2a + 1$, these can be multiplied to give the area, either in sections, as we see in Figure 4.4, or as a single statement.

Whilst the algebra connections described are beyond the primary remit on the whole, appreciating the links will help you to understand the foundations being laid by work on arrays.

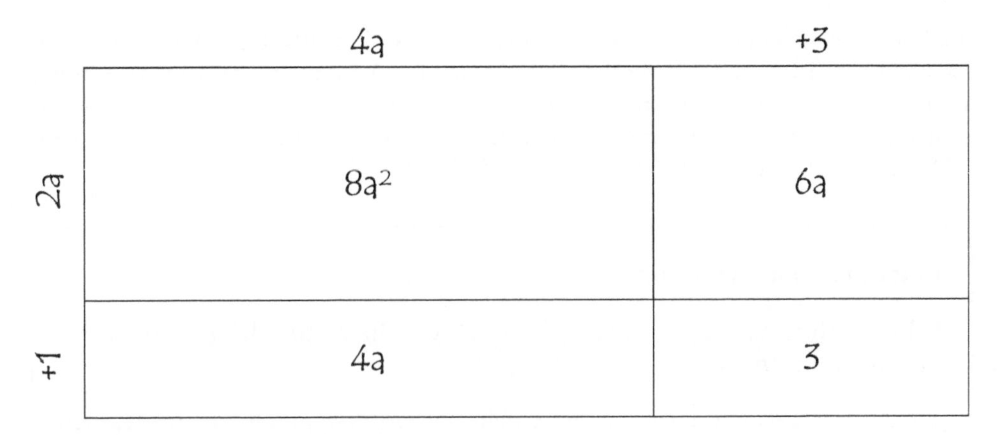

Figure 4.4 $8a^2 + 10a + 3$ shown using area as a model

Fundamental multiplication facts

As with addition and subtraction, mental methods should be considered a child's initial option if the numbers presented in a multiplication calculation lend themselves to a mental approach. It is clear that children need to be encouraged to learn their times tables, but equally important that children should have enough depth of understanding to be sure that they are recalling and applying the number facts correctly, and that they have the capability to work out any that they cannot remember. It was not particularly fashionable to learn multiplication facts when I was at school, and there are still some that I'm conscious of the need to check to be sure. Take 7×8; this I know to be 56 but only because I am really sure that $7 \times 7 = 49$, and the addition of one more seven takes me to 56.

Armed with predominantly instant recall of multiplication facts, learners can then employ these quickly and easily in the following ways:

- to multiply digits, for example, in long multiplication;
- to calculate related facts involving multiples of 10, 100, etc., for example, if $3 \times 5 = 15$, then $3 \times 50 = 150$;
- to calculate related facts involving decimals, for example, if $4 \times 2 = 8$, then $4 \times 0{\cdot}2 = 0{\cdot}8$;
- to build facts beyond the typically learned times tables, for example, the 13× table could be generated by combining the 10× and 3× tables.

Additional facts are gradually acquired by inquisitive and observant mathematicians, and clearly this can be promoted by effective teachers who are equally observant. Facts like realising that 25 is a quarter of 100 ($25 \times 4 = 100$) and so multiples of 25 can easily be related to multiples of 100.

Relationships between times tables are also powerful, and many such links rely on doubling and halving skills. If I can double, then I can master 2× almost any number;

doubling that result gives me 4× and doubling that 8×. Similarly, if I can work out what 3× something is, then doubling that will give me 6×. Multiplying by 10 is also pretty straightforward and if I halve my result I have achieved 5×. As multiplication and division are inverse operations, one can be used to reverse the other, thus if $6 \times 3 = 3 \times 6 = 18$ we know by default that $18 \div 3 = 6$ and $18 \div 6 = 3$.

Common misconception

'I know that when you multiply by 10 you have to add a zero to the end of the number, so 5·2 × 10 = 5·20'

Building factual knowledge sometimes falters where assumptions are made, such as the one exhibited here. The key teaching point is that if something is multiplied by 10 it will get 10 times bigger; the digits therefore shift one space to the left. In whole-number examples such as $74 \times 10 = 740$ the zero appears as we need to show the absence of units in the answer. In the case of $5{\cdot}2 \times 10$ the units will be filled by the 2 as it moves from the tenths position; we do not need to signal the now empty tenths column since there is nothing beyond it to the right and we also lose the decimal point, as a result giving us $5{\cdot}2 \times 10 = 52$.

The language, resources and methods of multiplication

Language and symbols

The everyday language of early multiplication should ideally stem from the real-life contexts advocated earlier in the chapter and will often just focus on the numbers involved, so two packets of sweets, with five per packet, becomes 'two fives', equalling ten. This then progresses to use of terms like 'times' and for the same example we could now say 'five times two' and record this using the symbol for multiplication: $5 \times 2 = 10$. Gradually modelling other language to read this number sentence, such as 'five multiplied by two' will help to develop children's awareness that the same calculation can be read in different ways.

As well as the language we use with children, there are also professional terms which help to describe the different roles the numbers play. In the example $5 \times 2 = 10$ the 5 is the 'multiplicand', the amount to be replicated; the 2 is referred to as the 'multiplier', the number of times the original amount is to be repeated; and the answer, 10, is referred to as the 'product'. When working in context it should be borne in mind that whereas the multiplicand will carry any units related to the context, the multiplier will not. The multiplier only tells us the extent to which the starting amount is to be replicated. So if I tell you I have three lengths of ribbon, each 24 cm, the calculation would be 24 cm × 3 = 72 cm to find out how much ribbon I have in total, albeit in three separate pieces. Note the absence of units for the multiplier. The multiplier can also be used to describe the scale factor in increase and decrease scenarios; if I buy another length of ribbon three times as long as my original piece, the calculation looks pretty much the same, but this time I have a single piece of ribbon. Measurement contexts work well for scaling, since something can take twice as long, be three times heavier and so on.

'Multiple' describes the numbers we get when we repeatedly add the same amount; typically positive whole numbers. If we count in threes, 3, 6, 9, 12, 15, ..., all these numbers are multiples of three. Familiarity with these number patterns stems from lots of repetition of them and is enhanced by drawing the children's attention to the position of the number in the pattern. In this example, 12 is the fourth number in the sequence and this helps us to appreciate that 3 × 4 = 12. For further commentary on number patterns, see activities 4.12–4.14.

Number system resources

At the start of the chapter we met the idea that early resources for multiplication benefit from being contextual and inherently multiplicative. Later resources begin to draw more on the structure of the number system.

Hundred squares, a common classroom resource, can be a very powerful image to show the spread of different multiples, including the fact that some numbers are multiples of many numbers while others fail to appear in any of the times tables other than × 1 because they are prime. The multiples can be noted in whatever way suits your classroom, such as circling numbers in different colours on the interactive whiteboard, shading a personal copy or laying different colour counters over or next to the numbers as we see in Figure 4.5.

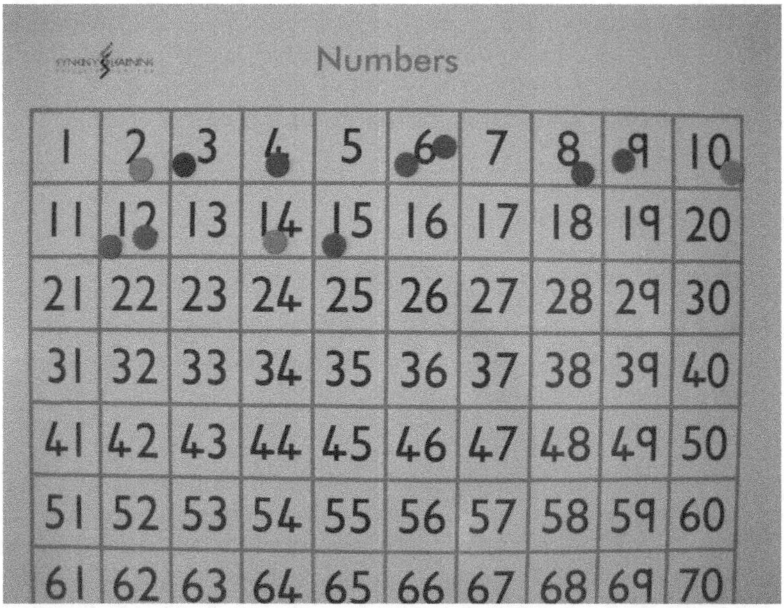

Figure 4.5 Hundred square showing the first few multiples of 2 and 3

The number line can be used to illustrate the idea of a repeated jump, and, like the 100 square, can illustrate relationships between times tables. In Figure 4.6 we see that the 3× and 5× tables will both land on the number 15. The distinction between multiples and counting on in threes or fives from any number should be noted; counting in fives from 2 gives us the pattern 2, 7, 12, 17, 22, 27, ..., none of which are multiples of 5 and which is therefore not about multiplication; in contrast the pattern 0, 5, 10, 15, ... gives the exact multiples of 5.

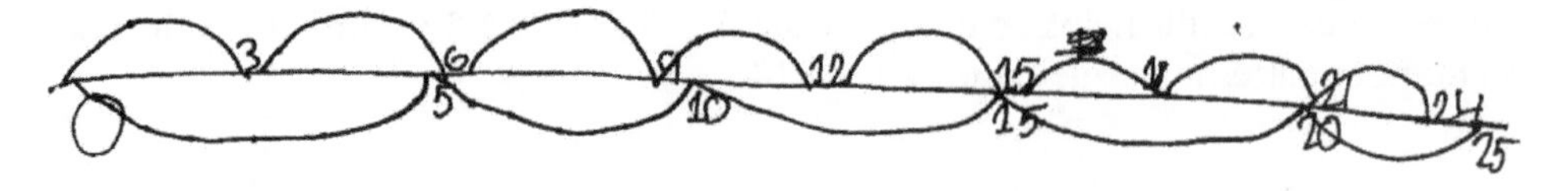

Figure 4.6 The 3× and 5× tables shown as jumps along a number line

As well as whole numbers, number lines are able to illustrate number patterns involving fractions and decimals, for example 0, 0·25, 0·5, 0·75, 1, 1·25, ... or $\frac{1}{3}, \frac{2}{3}, 1, 1\frac{1}{3}, 1\frac{2}{3}, 2, 2\frac{1}{3}, \ldots$ The notion of repeated addition and the associated number line imagery stop being helpful, however, in situations involving multiplication of two fractions or decimals. A calculation such as $\frac{2}{3} \times \frac{3}{4}$ does not lend itself to being represented or worked out in this way and alternative approaches are discussed below.

Working with fractions and decimals

In the primary sector one of the best ways of dealing with the multiplication of fractions is to try to relate it to a real-life context. To illustrate this let us imagine a long stick one of your children has picked up in the woods. The stick is the whole, so in the calculation $1 \times \frac{1}{4}$ start with the whole stick; multiplying this by $\frac{1}{4}$ has the same effect as dividing by 4 and we end up with just $\frac{1}{4}$ of the original stick. Alterations to the order and the language can help here, with $1 \times \frac{1}{4}$ becoming $\frac{1}{4} \times 1$ and being read as 'a quarter of one'.

Common misconception

'Multiplication always makes bigger answers'

Clearly this is not the case, not only in our stick example, where $1 \times \frac{1}{4} = \frac{1}{4}$, but also when multiplying by 1 (where the answer is the same as the multiplicand) or by 0 (where the answer will also be 0).

But what if both numbers were fractions? What does $\frac{1}{4} \times \frac{1}{2}$ say about the stick scenario? Well, if we start with only a quarter of the original stick and then take only half of that, we end up with just an eighth of our originally long stick. Given the commutative nature of multiplication it would not make any difference if we approached this the other way round, as seen in Figure 4.7. 'Half of a quarter' and 'quarter of a half' amount to the same thing!

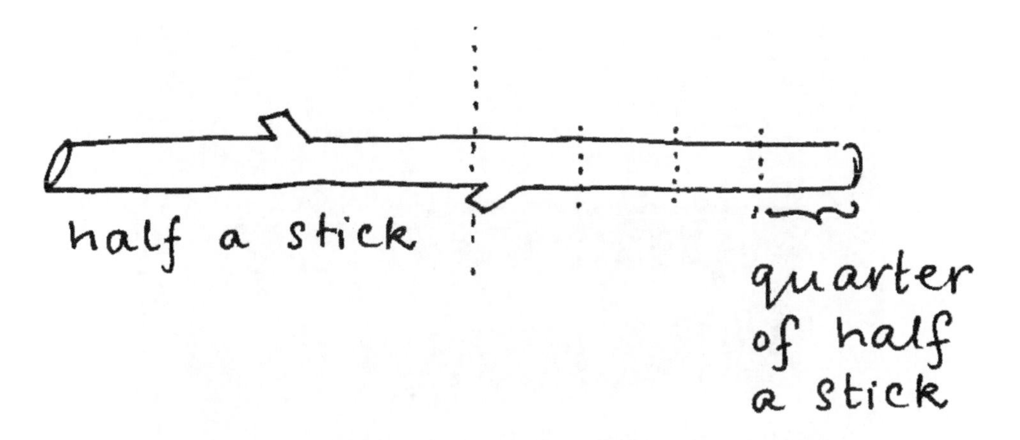

Figure 4.7 Stick image illustrating $\frac{1}{2} \times \frac{1}{4} = \frac{1}{8}$

Similarly with decimals, contexts can help to support children's understanding. If the decimals involved are couched in terms of money or measurement these can be made more accessible to children, noting that the units would only apply to the multiplicand and to the product; so the calculation of 3 weeks' worth of pocket money paid at a rate of £2·50 per week could be recorded as £2·50 × 3 = £7·50. Were we to ask for a pocket money advance, say, half-way through a week to enable a special purchase, then we would perhaps only get half the amount as reflected by the multiplier: £2·50 × 0·5 = £1·25. For more on multiplication of decimals, see the section below on written approaches to multiplication.

With both decimal and fraction multiplication consider the advantages of starting out with one of the numbers kept whole before going onto multiplication of two decimals or two fractions. Whole numbers provide a more accessible entry point, and returning to them may be necessary from time to time for exposition when children are struggling to multiply fractions and decimals.

Place value resources

Access to place value equipment is particularly useful for developing children's ability to multiply increasingly large numbers, noting the role of the distributive law in allowing us to split the tens from the units to make calculation easier. Start with simple examples such as 12 × 3 and use place value equipment to make a twelve (one ten and two units), another twelve and another twelve. Combine the parts to demonstrate the fact that there are now three tens and six units, so 36 in all as seen in Figure 4.8.

With slightly larger numbers it will be beneficial to start demonstrating multiplication by arranging apparatus as a rectangular array, showing however many rows and columns. This helps to lay foundations for written approaches to multiplication and is illustrated in Figure 4.9, which represents 16 columns and 12 rows but is 'built' using a 100 square, eight tens (80) and 12 units.

Figure 4.8 12 × 3 = 36 shown using place value equipment

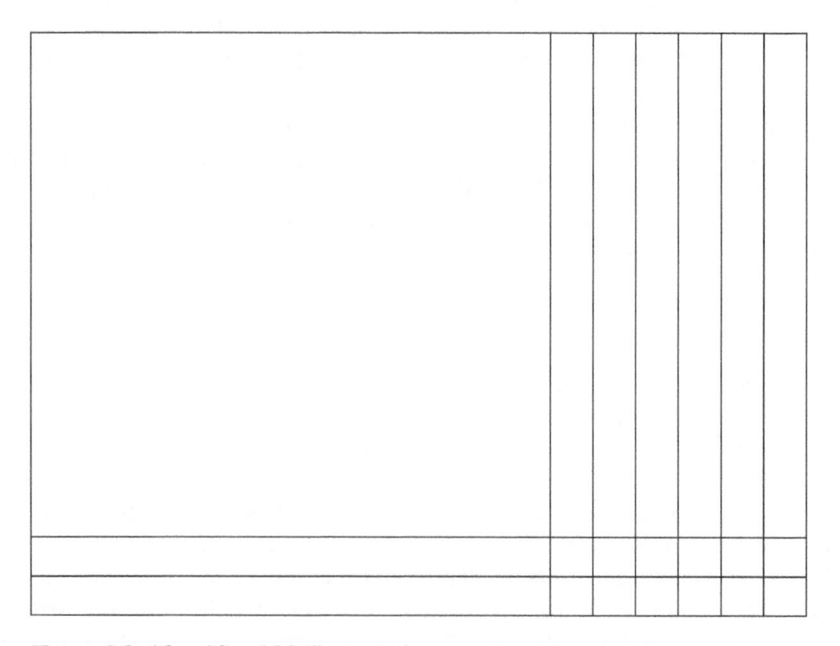

Figure 4.9 16 × 12 = 192 illustrated as a rectangular array

Written approaches to multiplication

Clearly, as numbers get larger, use of place value equipment for calculation becomes both unmanageable and inefficient, and children will progress to informal and then formal written methods. For more on informal methods for multiplication and other operations, try Haylock (2010) and read about what he refers to as 'ad hoc' methods.

One way of building upon the earlier foundations laid using tens and units materials is referred to as 'grid multiplication', an informal method for dealing with numbers by splitting them into their constituent parts. In Figures 4.9 and 4.10 we see the same calculation, 16 × 12, but in the latter example, rather than the sections representing the numbers involved, they have become spaces in which to record the answers, four in all. Calculating these relies on the child having secure knowledge of basic multiplication facts. Having got these four intermediate answers the final stage is to add them together to get the overall total, and, as an informal method, the way in which this is recorded need not be prescribed.

×	10	6
10	100	60
2	20	12

Figure 4.10 Grid multiplication showing 16 × 12 = 192 (added mentally)

One of the major advantages of grid multiplication is that the user can retain a sense of the size of the numbers involved. In the example in Figure 4.11 it is clear that both large and small numbers are involved, as we have thousands and decimals.

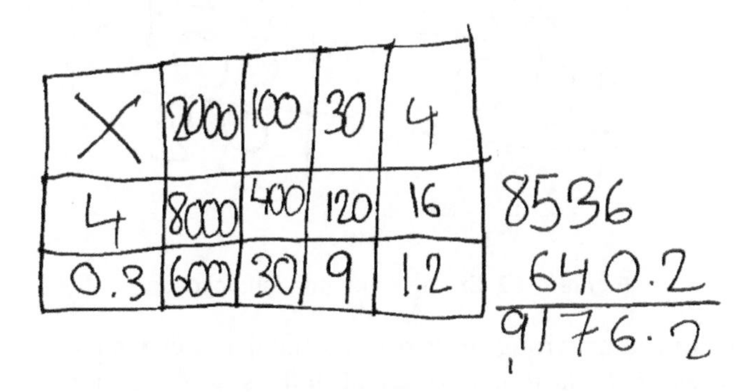

Figure 4.11 Grid multiplication showing 2134 × 4·3 = 9176·2

Once children have a really secure grasp of grid multiplication they tend to progress onto what is referred to as 'long multiplication', a formal written method for multiplication which reduces the number of steps, aiming to make the process more efficient. Much like other formal written methods, it tends to treat the numbers as a series of digits, starting from the units, rather unlike the more informal methods which typically focus on the numbers as a whole, starting with the largest values. In Figure 4.12 we see two more demonstrations of our earlier calculation 16 × 12 = 192, this time using long multiplication. The numbers have been swapped round in the example on the right, just to demonstrate a different path to same answer. The process started with the top line each time, multiplying first the unit number and then the tens value by the unit on the bottom line, giving the first of two answers. Note the possibility of variety: some people start by multiplying by the tens value and thus the two rows of answers would be the opposite way round. Whichever order you opt for, it is important to record a zero in the units position when multiplying by the tens value (in this case an actual 10); although we are still treating the 1 as a unit for the purpose of multiplying, our answer will be ten times bigger. The final stage is to add the two answers to give the total, hence secure column addition is a prerequisite for long multiplication, as well as sound times table facts.

In Figure 4.12 we naturally have the same answers for 16 × 12 as we saw in Figure 4.10 when we were using grid multiplication. Compare the different pairs of numbers generated in each of the long multiplication examples with the four answers generated through the grid multiplication process. You may have spotted the way in which some of those answers have combined to give the values we see in Figure 4.12 as a result of long multiplication. This provides the basis of activity 4.11 where the children have the opportunity to explore these connections.

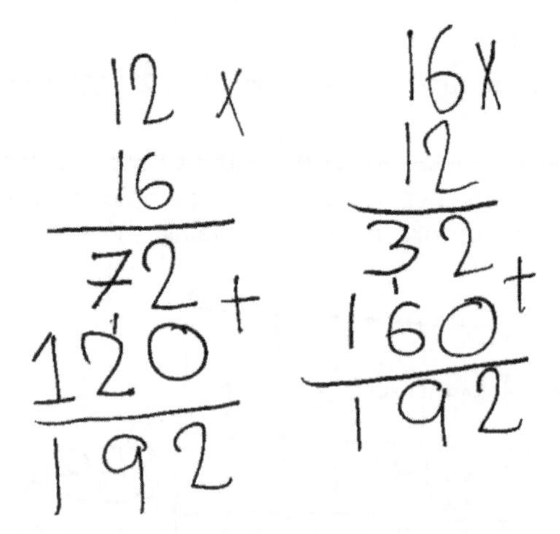

Figure 4.12 Examples of long multiplication

If children are getting wrong answers it may be that they would benefit from further work with grid multiplication to re-establish the sizes of the numbers involved as well as seeing each different part of the process separately.

Common misconception

'$57 \times 34 = 43$ using long multiplication'

Wrong answers can be very telling, and two possible errors are combined here. The child has probably treated all the digits as if they were units since $7 \times 4 = 28$ and $5 \times 3 = 15$ and $28 + 15 = 43$. Correcting that misconception will involve realising that rather than 5×3 this calculation is actually 50×30, so we are definitely expecting an answer in excess of 1500; hopefully the child would quickly realise that theirs could not possibly be the right answer!

The second error relates to the 'lost' parts of the calculation: 7×4 and 50×30 are just two of the four calculations required, and the ones we are therefore missing here are 50×4 and 7×30. Whereas in grid multiplication we see all four parts separately, they are combined in long multiplication for efficiency and this does sometimes result in errors.

Whichever method a child uses, it is important that they can use the method accurately and with confidence. In Chapter 3 we explored the idea of approaches to calculation being ideally accurate, efficient and flexible, and of course this applies to multiplication just as much as it did to addition and subtraction, with all the attendant arguments. James, a competent mathematician, once said to me that he preferred grid multiplication to long multiplication, but in Year 6 at the time recognised that he would be expected to use the formal written method in test situations. This is surely the mark of a good mathematician and a good mathematics education, that a child has a choice of method, albeit that the choice here was not being dictated by the numbers involved in the calculation, but by the testing regime!

Long multiplication can clearly be used for multiplication of numbers beyond tens and units, as well as of decimal values. In this scenario we can essentially ignore the decimal point at the calculation stage, remembering to put it back in at the end of course! But where to put it? Practising the skill of estimation will pretty much ensure I do not put a decimal point in the wrong place. Long multiplication of 639×24 gives me an answer of 15,336, so I know that $6{\cdot}39 \times 2{\cdot}4 = 15{\cdot}336$ because 153·36 would be far too large and 1·5336 too small for what is essentially 'six and a bit times two and a bit'. Because $6 \times 2 = 12$ and $7 \times 3 = 21$ my answer has to fall somewhere between these two limits. The other method for replacing the decimal point when multiplying is to count the number of decimal parts; having started with three altogether in the original calculation, our answer should be the same. I need to sound a note of caution here, however: in some answers the decimal point and any decimal parts will not be obvious. Take the calculation $1{\cdot}4 \times 5$ for example. The answer is 7 and there is not a decimal point in sight. As $14 \times 5 = 70$, putting the decimal point in renders both the zero and the decimal point unnecessary since we would not typically give an answer of 7·0 showing the decimal point and absence of tenths.

Clearly, whatever method children choose to use for whatever numbers they are seeking to multiply, they should be encouraged to at least have an idea of what sort of answer to expect, for example, through making a rough estimate. Working with children to ensure they get into the habit of this means they are more easily alerted to possible errors, and ensures we are doing all we can to produce thinking mathematicians!

Activities for multiplication teaching

As with all the other chapters, the activities described are merely suggestions and you should feel free to adapt them to meet the needs and interests of your own class in different parts of the lesson. Remember that many benefits are obtained through encouraging the children to communicate mathematically with each other, for example, to share ideas for approaches to calculation.

Pupil activity 4.1

Minibeasts for multiplication
Y1 to Y4

Learning objective: To develop conceptual understanding of multiplication.

The fact that creepy crawlies have such a variety of numbers of legs really lends itself well to the topic of multiplication. Take woodlice; they typically have 14 legs, so how many legs in total if there are three of them scuttling around?

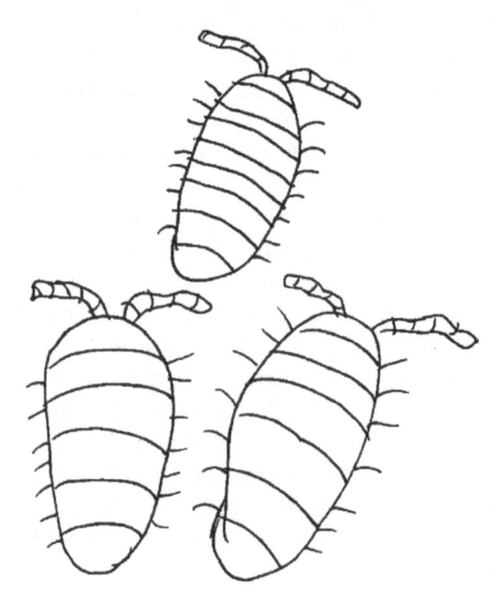

Having introduced the children to the idea and through other work ensured they have enough knowledge regarding numbers of legs, give them some freedom to explore creatures of their choice. Whilst the children are initially likely to draw pictures to support the calculation of answers (by counting numbers of legs), try to gradually encourage them to think about answers they can work out without having to draw and count. Some of the calculations some of the children attempt may surprise you, and of course work can be recorded in a range of ways from pictorial and informal to use of recognised symbols ($14 \times 3 = 42$ or $14 + 14 + 14 = 42$). Given the open-ended nature of the activity, you will be afforded good assessment opportunities: of the children's conceptual understanding of multiplication and of their approaches to recording. Having gathered such information you can then build on it, for example, by reintroducing the symbol for multiplication if the children are predominantly using repeated addition. Extreme examples can help to prompt this; if I have 100 butterflies I might well know the answer without drawing, but probably do not want to write $6 + 6 + 6 + 6 + \ldots$!

If there are particular times tables you want to explore but no suitable minibeasts then be creative: who has not met a seven-legged spider who has had a nasty plughole accident? What if there were three of them? I would also strongly recommend including examples which force the issue of multiplication by 0 (worms are ideal; however many I dig up in my garden there will still be no legs!) and by 1 (such as snails with their single 'foot').

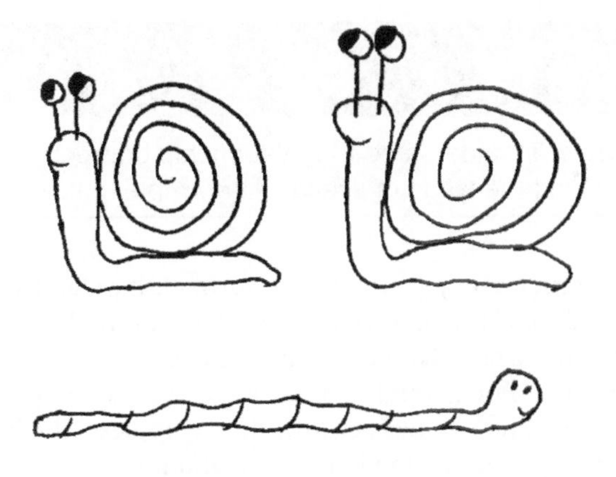

Pupil activity 4.2

'I'm hungry!' Rehearsal of multiplication facts
Y1 to Y5

Learning objective: To develop recall of multiplication facts in context.

Developing a really secure conceptual understanding of multiplication is aided by use of contexts, since these can help to build mental images. Continued use of the stories and imagery is then a fun way of rehearsing multiplication facts, hopefully to the point where many of them are on automatic recall. I have chosen eating as the context here, not only because it is one of my favourite occupations, but also because it is a topic children can relate to. The banter focuses on starting with a particular number of strawberries in a bowl, or biscuits in a packet, but being so hungry that you are going to eat double, or five times as many or maybe even a hundred times! Rather than avoiding big numbers, try to include them as par for the course; consider also opportunities to exemplify multiplication by 0 and 1. For example, fancying the bowls of strawberries but finding that all of them are empty!

In the early stages you may wish to focus mainly on ×2 through eating double amounts, but you will be able to judge the children's readiness for other times tables through questioning. Where children cannot recall the facts and are struggling to work them out mentally, consider whether drawing pictures or modelling with resources would support thinking. Note that this is a temporary solution since recall is the long-term aim.

Clearly both context and numbers involved can be altered, but aim to have fun with it – relaxed mathematicians will remember more!

Pupil activity 4.3

Multiplication stories
Y1 to Y6

Learning objective: To understand the relationship between written multiplication calculations and the types of scenario they represent.

In activity 3.5 we considered opportunities to turn addition and subtraction calculations into stories, and we can of course do the same with multiplication. This can be as simple as turning over a calculation card and a picture card and thinking about how the two might be related. For example, if I turn over a card saying 5 × 2 and a picture of an apple I might imagine two bags each with five apples. Children could be asked to respond by drawing what they think the calculation might represent, perhaps with a one-minute time limit since understanding of multiplication is the intended learning outcome and not beautiful art work! This will be a particularly good way of assessing the child's ability to think multiplicatively; were they to draw one group of 5 apples and another of 2 apples in response to the above scenario it would be clear that they had not fully grasped the concepts involved in multiplication or were not secure in their interpretation of the symbol '×'.

Note that where stories relate to measures it is important to ensure that the appropriate numbers have units attached, therefore with older children we might specifically use calculations which model this. Rather than using the picture cards the children would be expected to use the specified units to guide them as to appropriate scenarios, described in words or using a combination of words and pictures. What might they suggest for 5 ml × 2 = 10 ml or for 150 g × 2 = 300 g? The first could realistically be two 5 ml spoonfuls of medicine and an activity such as this one will help you to assess the children's understanding of measures as well how multiplication is recorded. Clearly the numbers and units could be varied depending on the ages and abilities of the children and you might want the children to supply their own answers!

Pupil activity 4.4

Measures and multiplication
Y3 to Y6

Learning objective: To understand the concept of multiplication in the context of measures.

Using packaging is ideal for exploration of 3D shape (see Chapter 10) but also lends itself to multiplication because of the measures information packets tend to display. The suggestions below can of course be adapted to involve other aspects of measurement and made easier or harder by changing the numbers involved.

- My favourite biscuits weigh 400 g and as they are on special offer I think I'll stock up and buy eight packets. How heavy will my shopping be?
- The biscuits are currently 87p. How much will my shopping cost me?
- Because I might also get thirsty I think I'll get some cans of fizzy drink. Each can contains 330 ml. Is that more than 2 litres if I buy a six-can pack?

Note these contexts are more challenging than the minibeasts ideas presented in activity 4.1 not only because some of the numbers here are more complicated, but also because the legs of the minibeasts could be counted whereas here counting is not possible in the same way.

Pupil activity 4.5

Estimation linked to multiplication
Y1 to Y6

Learning objective: To develop awareness of the role of estimation in multiplication.

Early estimation of answers to multiplication questions will ideally be contextual; a child has a look and thinks that four tricycles will have about ten wheels and then counts to check and finds it's actually 12. Gradually more advanced multiplication strategies are used to check the original estimate.

As calculations get more demanding, other aspects of estimation come into play, in particular the idea of taking some nearby numbers that will prove quick and easy to multiply to provide a rough estimate. For example, I know that 20 × 20 = 400 so I can predict 19 × 21 is going to be somewhere around the same answer. To explore this further, each small group can be given a range of calculations presented on cards, as shown below, to sort into piles which they think will give similar answers. Aside from 50 × 5, note that they are not easy calculations; they need to be challenging enough to encourage the children to apply the estimation skills of working out nearby calculations which are simpler. The decimal has been included to see whether the children can apply their knowledge to non-integer numbers. Which of the following examples do you estimate would have similar answers and how do you know?

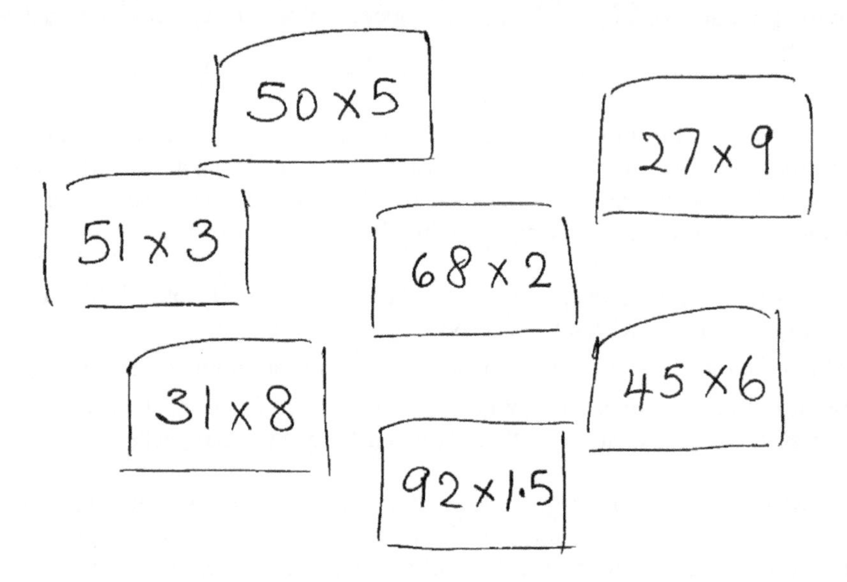

Once the children are sure they are happy with their decisions, they can use a calculator to investigate just how close the answers in each of their piles would be. Encourage groups to share any surprises and to try to articulate what they think caught them out. The extension for this task is to use estimation skills to predict the actual order of the cards before resorting to the calculator, and the more children engage in activities like this, the more likely they are to begin to come up with actual answers based on their attempts to estimate (for more on this, see activity 4.10 on adjustment strategies).

When estimating answers for multiplication calculations involving decimals, the idea of limits is a useful one. The limits are obtained by rounding each number down to the nearest whole number and multiplying to give the lower limit, and rounding up to the next whole number and multiplying to find the upper limit. Therefore 3·5 × 2·4 must be greater than 6 (as 3 × 2 = 6) but smaller than 12 (as 4 × 3 = 12). As we see on the next page, the child has determined the upper and lower limits using this approach, selecting the calculations from a range displayed on the board.

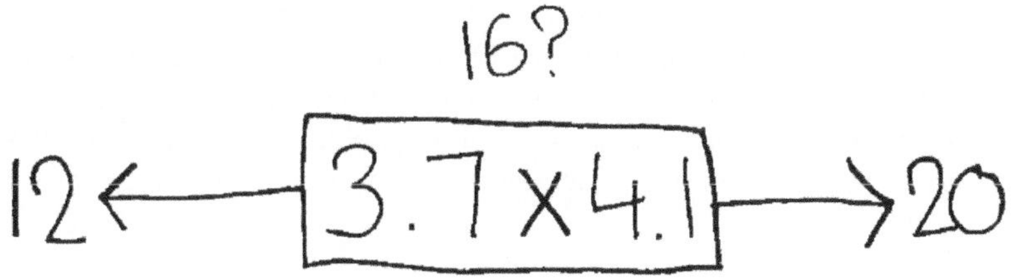

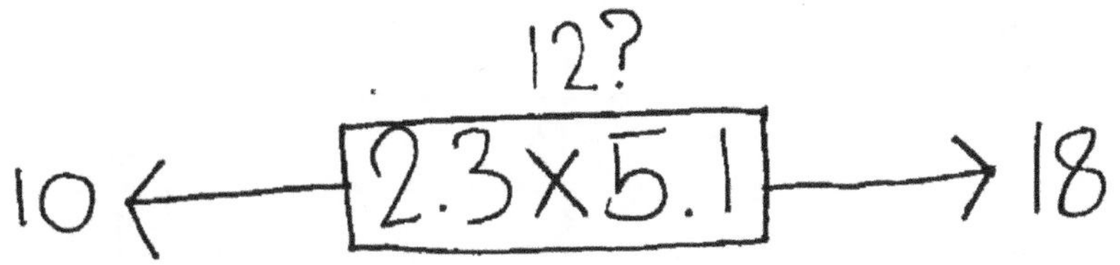

16?

12 ← 3.7 × 4.1 → 20

12?

10 ← 2.3 × 5.1 → 18

In discussion the child was also asked to think about which limit each calculation was closest to and why they thought that; he knew, for example, that 2·3 × 5·1 would be closer to 10 as the number of tenths was quite small and has added his 'best guess' in both cases, the number with the question mark. This focus can of course be incorporated into the activity and contributes to determining more accurate limits as children get older.

Pupil activity 4.6

Sets of multiplication (and division) facts Y2 to Y6

Learning objective: Identification of relationships between numbers in both multiplication and division.

3 15 5

Given these three numbers (or similar), can the children identify any relationships between them? Hopefully they will spot that as well as 3 × 5 = 15 and 5 × 3 = 15 we can also have 15 ÷ 3 = 5 and 15 ÷ 5 = 3. Depending on the age of the children, these facts may not involve formal recording using symbols as seen here, and can be modelled using resources if you feel that will help to cement the understanding of some children. The numbers chosen can of course be tailored to suit the ages and abilities of the children. I would envisage collaboration between children in finding the sets of number facts, as discussion helps to correct misconceptions and affords use of mathematical language.

The collections of numbers seen below may seem slightly bizarre; what answers would you give? Note that some of the sets of facts would be more limited, such as the first and last ones where there will only be two rather than the usual four. And what about division by zero, for example 1 ÷ 0? Division by zero is generally understood to have no meaning since the inverse is not possible: no number multiplied by zero will return us to our starting number! Red herrings, including the third line which does not work, are great for promoting discussion, so aim to include some such examples.

4	16	4
0	1	0
7	38	6
1	1	1

As well as providing children with sets of numbers to investigate, they could be asked to generate their own, or to work from pairs of numbers, such as 3 and 12. This makes the investigation far more open-ended since this pair could result in 3 × 4 = 12 and so on, or could involve 36 as the third number (3 × 12 = 36). With older children it could even progress to giving answers like $12 \times \frac{1}{4} = 3$ and $3 \div \frac{1}{4} = 12$!

Pupil activity 4.7

Building on multiplication facts
Y3 to Y6

Learning objective: Ability to use the most straightforward multiplication facts to generate others.

Whilst this activity revolves around the 17× and 19× tables, the same teaching approach can be related to the learning of simpler times tables and to other numbers too. It is about learning to think like a mathematician, someone who makes connections wherever possible.

If we were to ask our class who knows the 17× table, we would either get looks of disbelief, or a child or children confident enough to say 'yes'. Ideally it would be the latter – children with a predisposition to being able to generate new mathematics! Let us think about the order in which we might tackle working out what the multiples of 17 might be:

- First we might well consider 17 × 1 and 17 × 10; hopefully pretty easy facts.
- The next fact I think I would go for would be 17 × 5, as I like halving as a strategy.
- I like doubling too. So 17 × 2 is easy to fill in and from that I can generate 17 × 4 and 17 × 8.

- As 17 × 6 is also an even multiple it must be half-way between 17 × 4 and 17 × 8, but equally it's three times 17 × 2, and also we could add one 17 to 17 × 5. Having considered all the options in the light of the numbers involved, I have decided to go for adding 17 to the fifth multiple. And while I'm there I might as well add another 17 to get 17 × 7.
- Rather than addition I think I'll use subtraction to find 17 × 9 as it will be one less 17 than 170.
- Finally, let's consider 17 × 3; how about we add 17 and 34 since (17 × 1) + (17 × 2) is just what we are after.

17	34	51	68	85	102	119	136	153	170

Whilst the 17× table has been presented here in tabular form, labelling a counting stick with the multiples is another good way of involving the children in building the facts. A web of links is also a strong visual image and a good way of highlighting the connections and is demonstrated here using the beginnings of the 19× table.

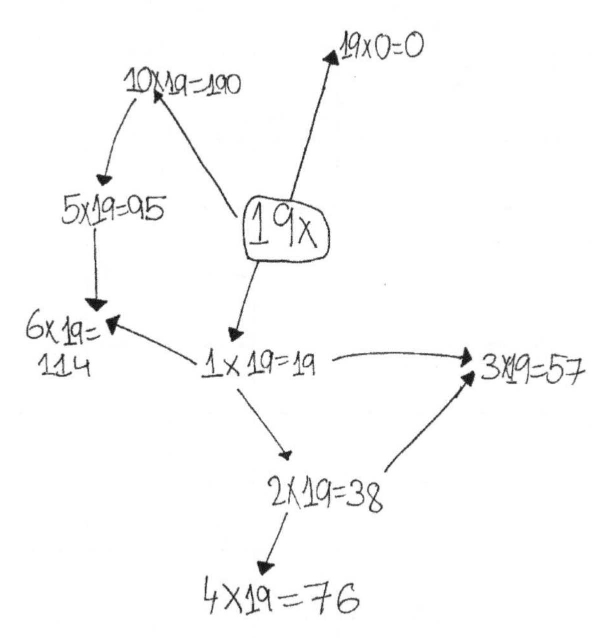

Pupil activity 4.8

Multiplication grids
Y3 to Y6

Learning objective: To build an in-depth understanding of the information presented in multiplication grids.

Whilst children are regularly set tasks like completing multiplication grids, I suspect these tasks could often be better used by asking children to attend more closely to some of the features. The questions below suggest some things you might direct the children to think about:

- If you had to complete as much as you could in a limited time, where would you start to maximise your chances of getting lots filled in?
- Which numbers appear twice and why is this?
- Do any numbers appear more than twice? Which numbers appear the most?
- Are there any answers which appear an odd number of times in the grid?
- Will any numbers appear only once in the grid and, if so, why?
- What about prime numbers? Are there any?
- Would the answers to these questions be any different if we extended the grid beyond 10×10?

X	1	2	3	4	5	6	7	8	9	10
1										
2										
3										
4										
5										
6										
7										
8										
9										
10										

You could no doubt add questions of your own, anything to promote the children's more active engagement with the numbers in the grid. Just filling in such a grid might help with a child's recall of multiplication facts, but drawing their attention to those facts more explicitly will help to ensure this is the case.

Pupil activity 4.9

Tens and units multiplication
Y2 to Y4

Learning objective: Ability to multiply a teens number by a single digit with the support of base 10 equipment.

In the phase where early multiplicative concepts are being established, children may rely heavily on counting and adding strategies to answer multiplication questions, but this becomes impractical as the numbers get larger. As we saw earlier in this chapter, place value equipment can be used to bridge the gap between this early work and formal written methods. Once children are secure as to how numbers are represented using base 10 apparatus, you can therefore begin to introduce them to the idea of having several of the same amount, rearranging the cubes to make it clear that there are however many tens and so many units. A simple way of having children generate calculations of this type is to have all the numbers from 11 to 19 (and possibly beyond if you have plenty of base 10 equipment) written on cards and placed face down; choosing a card gives the child the multiplicand or size of group and then a dice is rolled to give the multiplier. Whilst many children will require the support of the apparatus, some may progress quickly to working entirely without it or to sketching it quickly as an informal paper and pencil aid.

Of course the type of dice can be varied to assist with differentiation as well as choosing particular sets of number cards, depending on the ability of the children. Larger numbers may prompt some children to move away from reliance on the equipment, such as to calculate 53 × 3, and the occasional really big number will definitely force the issue; 11,000 × 2 may not be possible using the place value apparatus you have in your classroom!

Note that whilst these activities are aimed at the younger primary children, it may be beneficial to revisit the use of place value equipment with older children if you find that some of them are struggling. This tends to occur at key points such as the introduction of vertical written methods for multiplication, and demonstrating the relationship between the cubes and the written method may provide the solution to a smooth transition.

Pupil activity 4.10

Adjustment strategies for multiplication
Y3 to Y6

Learning objective: To recognise opportunities to use adjustment strategies for mental multiplication and begin to appreciate that care is needed as to whether the adjustment should be made by addition or subtraction.

As we saw in activity 4.5, ability to estimate answers to multiplication questions relies on understanding which easy calculations are nearby and can therefore be used to give a rough idea of the expected answer. In many ways adjustment strategies use the same approach, taking it one step further to actually complete the calculation based on the easier starting point. For example, 43×19 is close to 43×20 (which is somewhat easier, especially if I realise that I can interpret this as 43×10 twice, which equals 860). The answer can then be adjusted to reflect that I should only have calculated 43×19, therefore $43 \times 19 = (43 \times 20) - (43 \times 1) = 860 - 43 = 817$. Here the adjustment relies on subtraction to arrive at the right answer; for 43×21 we might see addition in use if the approach taken is $(43 \times 20) + (43 \times 1) = 860 + 43 = 903$. Note that there is no necessity to record what are, essentially, mental calculations; whilst I have written them formally here to illustrate my points, in practice I would only jot down any numbers I could not hold in my head. The role of adjustment strategies is to support mental multiplication and informal written approaches.

Common misconception

'$7 \times 19 = 147$ as $7 \times 20 = 140$ and I have adjusted my answer'

Adjustment strategies do not seem to work particularly well for some children as they have a tendency to make the adjustment in the wrong direction.

This activity gives children the opportunity to explore adjustment strategies in a carefully structured way using money as the context. The reason it works well is because children can visualise the coins to help them. Have a selection of something like biscuit bars and model wanting to buy several, say, four, at a cost of 20p each, literally allocating a 20p coin per bar. But what if the bars had only been 19p? Can the children suggest a way of working out how much I will need to pay now? Perhaps we could still give the shopkeeper the four 20p coins and get 1p change on each of them; effectively we have paid $(20\text{p} \times 4) - (1\text{p} \times 4) = 80\text{p} - 4\text{p} = 76\text{p}$, but the important aspect of the teaching is the discussion of ideas for working this out and not the ability to record it. To reinforce the fact that we will pay less if the bars are cheaper, I recommend exploring the opposite scenario: the shopkeeper has put prices up and bars are now 21p, so how does that alter how much I have to pay? Model various examples around 10p, 50p and £1 before letting the children explore some of their own, perhaps choosing from a range of paired calculations since the pairings will help to focus the children's attention on the fact that some answers will be adjusted upwards and others down. A couple of examples are given below:

How much do I need to pay?	
53p x 4 =	47p x 4 =
Explain how you worked out your answers...	

How much do I need to pay?	
98p x 7 =	£1·02 x 7 =
Explain how you worked out your answers...	

Obviously as children get more experienced in their use of adjustment strategies they can begin to look for other opportunities to employ them, including other contexts as well as for context-free calculations. Note that some calculations lend themselves to adjustment strategies less well; your awareness of those for which adjustment strategies would work well (typically numbers close to a multiple of 10) will ensure that you include such examples in ranges of calculations presented to the children. Older children will hopefully require less explicit modelling but will still benefit from reminders that adjustment strategies can make a mathematician's life easier: £1.99 × 8 may not seem very easy but is close to £2 × 8 which feels much nicer!

Pupil activity 4.11

Comparing written methods for multiplication
Y3 to Y6

Learning objective: Awareness of the similarities and differences in how written methods for multiplication work.

For children to use any written calculation method reliably they really need a deep understanding of how it works, and one of the best ways of achieving this is to compare different methods looking for similarities and differences. Rather than being about getting through copious amounts of work (numerous calculations), it is about spending quality time really picking apart just a few examples, examples which are initially presented completed since the emphasis is on discussion of the whole process.

```
   58 ×
   24
  ----
   232
  1160 +
  ----
  1392
```

×	50	8
20	1000	160
4	200	32

So 58 × 24 = 1392

Given that the answers are the same, encourage the children to spot the way in which the same numbers appear as each method progresses, albeit combined values in long multiplication.

Rather than always comparing different methods, bear in mind that the children could be asked to look at different versions of the same method, such as the numbers presented the opposite way round, or split in a different way as we see below. Here I have chosen not to show the answers as the children may need to satisfy themselves that the different split will not affect the answer.

×	20	4
20	400	80
5	100	20

×	12	12
25	300	300

Our aim, remember, is depth of understanding and not the ability to follow algorithms without really having to think, so after discussion of some examples I would give the children some calculations to try, asking them to present each one in at least two different ways. Pairs of numbers could be chosen by the children from a selection, picked from raffle tickets or whatever.

In the early stages of teaching about written methods consider using a similar task, but this time comparing the recording of a calculation with the same calculation modelled using place value equipment (such as we saw in activity 4.9).

Pupil activity 4.12

Multiple familiarity
Y1 to Y4

Learning objective: Develop familiarity with multiples through repeated exposure to them.

You will no doubt have your own innovative ideas for ensuring children experience multiples on a regular basis, such as one I came across years ago where the children performed brain gym activities whilst chanting the number patterns generated by various times tables!

For this activity the children need to be seated in a circle and the multiple you want to focus on is chosen in advance; say, multiples of 3, for the purpose of illustration. The class start counting round the circle in ones (always ones) with each child having to decide whether their number is a multiple of 3; if it is they have to stand up which of course will result in every third person standing once you have been all the way round. I rather like to carry on round the circle as different things happen depending on how many people are in class that day and whether it is a multiple of 3. If there are 30 people in the circle then the same person who stood up for '3' will remain standing for '33' and so on. If, however, there are only 29 people then something rather different happens: try it and see!

This activity has strong links with division and could be used successfully to draw the children's attention to factors and divisibility.

Pupil activity 4.13

Repeated addition on a calculator
Y1 to Y6

Learning objective: To develop ability to anticipate multiples beyond ×12.

Whilst the use of calculators might not be considered terribly fashionable in primary schools, it would be a real pity if their potential as a teaching aid were overlooked. With most basic calculators it is possible to make them add the same amount repeatedly by pressing a number and then + twice (sometimes once is enough) and then as you keep pressing = the calculator counts in multiples of your chosen number. So with 7++ and then = over and again we could generate:

7, 14, 21, 28, 35, 42, 49, 56, 63, 70, 77, 84, 91, 98,....

The value in playing around with this might relate to familiarity with the pattern (including going beyond 12×) and how the digits change, but a particularly nice focus is to ask the children to predict whether they will reach particular numbers as their pattern grows. Do you think I will get to 200 if I keep going in sevens? Why? Why not? Having thought about this in advance, the children can be asked to justify their reasoning, for example, the child who said they knew counting in threes would not 'hit' 100 as they thought they would get to 99 and then go past 100; they were right of course!

Most calculators will also generate patterns using decimals, and rather than addition the focus can be altered to repeated subtraction.

Pupil activity 4.14

Odd and even numbers and multiplication
Y2 to Y3

Learning objective: To appreciate that multiples of odd and even numbers each generate different patterns.

At first glance we might assume that odd and even numbers have very little to do with multiplication, but in fact the patterns generated when counting in different multiples are of only two types: one which is permanently even, and the other alternately odd and even. The idea of this activity is for the children to investigate different number patterns and to begin to articulate why this is the case. You might want to suggest a range of numbers from which children can choose numbers to explore; aim for a mix of regular times tables and larger numbers, ensuring both odd and even numbers are included. The choice does not really matter since the focus is on the patterns generated, and investigating in pairs or small groups will help to engender discussion.

Pupil activity 4.15

Digit patterns
Y3 to Y6

Learning objective: To develop understanding of number patterns created by the different times tables.

In addition to patterns generated by multiples of odd and even numbers, all multiplication patterns also have some order to the way in which the digits change, focusing in particular on the units. Some patterns are really obvious such as the units generated by the 5× table (5, 10, 15, 20), whereas for others children will have to explore much further ahead and be far more observant if they are going to spot the sorts of patterns that begin to occur with regard to the digits in the units column. Exploring tables such as 3× and 7× will help to make this point; their units' patterns actually run in the opposite direction to each other and exploring why this is the case would be a super question for able mathematicians to ponder! Having had time to explore, children can be asked to describe what they notice about their patterns, for example that the units in the 4× table go 4, 8, 2, 6, 0, 4, … as we see in the grid below.

4	8	12	16	20	24	

One of the side benefits from such an investigation is that children may begin to spot other relationships. For example, a child continuing the pattern for the 4× table might begin to identify with the blocks of 20 (4 × 5) after which the units' pattern starts again. This highlights the fact that 40 will be in the 4× table (4 × 10) as will 60 (4 × 15) and 80 and so on. Now try the 14× table: what similarities do you notice? What number does the units' pattern start to repeat at this time? Good mathematicians are observant; encourage the children to acquire new knowledge in this way!

Following lots of freedom to explore different times tables, a good way of developing this is to set a scene as follows:

There were four children and each chose a different times table (up to 10x) to investigate, writing down the numbers generated. This is what they noticed about their patterns:	
Child A	My pattern has no odd numbers.
Child B	I don't have the number 12 in my pattern.
Child C	My answers all have the same number of units.
Child D	All the digits from 0–9 appear in my units.
Which times table do you think each child was investigating? Why?	

Working collaboratively will ensure that children can discuss and challenge each other's ideas, for example, a child who thinks Child C can only have been investigating the 10× table but whose friend suggests maybe it was the zero times table! Aside from Child C there are clearly several possibilities for the child's choice of times tables, and this lack of certainty should therefore feature in the discussions.

A better understanding of digit patterns pays dividends when we get to tests for divisibility (see Chapter 5).

Pupil activity 4.16

Rectangular numbers
Y3 to Y5

Learning objective: To appreciate that all multiplication facts can be represented by at least one rectangular array, and that some of those rectangles will be squares.

This activity gives children the opportunity to investigate multiplication facts through rectangular arrays (on square paper or using arrangements of counters, for example), thinking about the range of facts and therefore the different rectangles it is possible to make. Working in pairs or small groups will support the children in finding all possible facts/arrangements, and dialogue promoting awareness of the differences between the numbers is important.

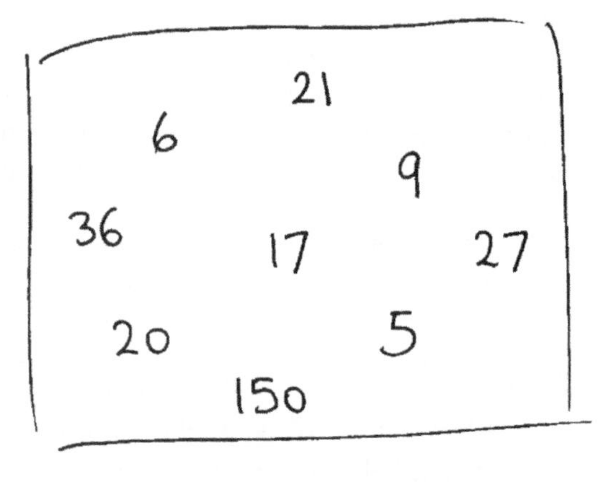

The numbers chosen here have been carefully selected to illustrate numbers which only make long thin rectangles (in other words, they must be prime numbers, such as 17); those which correspond to few multiplication facts (21); and rich numbers such as 36 (lots of multiplication facts but also a square number). A larger number (150) has been included since this may prompt children to draw on mental skills and known facts rather than physically trying to create an array. How many different rectangular arrays do you think there would be?

Note the links between this activity and activity 7.5 on area and perimeter relationships.

Pupil activity 4.17

Multiplication and the laws of arithmetic
Y2 to Y4

Learning objective: To secure the understanding that multiplication is both commutative and associative.

The rectangular arrays generated in activity 4.16 help to demonstrate the commutative law; that $21 = 7 \times 3 = 3 \times 7$, for example, where swapping the numbers over has no effect. These two activities continue this theme, multiplying more than two numbers and making links with the associative law. The second activity builds on the first.

- Children (either individually or in pairs) roll three dice, recording the calculation and the answer obtained by multiplying the three numbers. They repeat this quite a number of times before being asked to stop and have a good look at their own work. Are there any calculations involving exactly the same numbers? Were the numbers recorded in the same order? Hopefully the answers are the same! Widen the comparisons across groups or the whole class and stress the implications: that the order in which the same three numbers are written down and calculated makes no difference to the answer in multiplication.
- The teacher rolls three dice and positions them in different orders, calculating in the given order each time to remind the children that the answer will remain the same. The children then investigate, ideally collaboratively, whether the same answers can be obtained using other combinations of dice. A quick check suggests this is possible, as $3 \times 3 \times 4 = 36$ and $2 \times 3 \times 6 = 36$ as well. How many others can you find?

Of course it is possible to adapt both of these activities by using more dice or alternative dice to the regular ones. Consider using a dice with a zero since this will remind the children of the particular effect of multiplying by zero.

Pupil activity 4.18

Bit by bit multiplication
Y4 to Y6

Learning objective: To understand the role of the distributive law in multiplying several-digit numbers.

The distributive law is essential in multiplication since it underpins all written methods. This activity assumes children already have a reasonably secure grasp of what happens when you multiply by multiples of 10 and links to and builds on previous activities such as activity 4.9, where place value equipment was used to support multiplication of numbers in the teens. Whilst it does not help with the mechanics as such, it can be used alongside other work on calculation procedures to support the children in understanding how the numbers are partitioned as part of the written method. Children are given various calculations and have to find the cards which would match the calculation broken down into its constituent parts. A set of cards is illustrated overleaf, designed to match the partitioning we would see if we tackled these calculations using a vertical written layout.

14 x 6	10 x 6	4 x 6
57 x 8	50 x 8	7 x 8
35 x 7	30 x 7	5 x 7
34 x 27	34 x 20	34 x 7
28 x 41	28 x 40	28 x 1

The children could then be asked to convert the cards into written calculations, probably needing more support with the latter two. Here we would need to think about which products are obtained depending on which way round the two numbers are presented vertically. I chose 34 × 20 (680) and 34 × 7 (238) for 34 × 27 but it could equally have been 27 × 30 (810) and 27 × 4 (108), both of which give totals of 918. There would therefore be some mileage in preparing cards illustrating both and adding those into the mix. Plenty of scope for mathematical discussion in classrooms where such opportunities are welcomed!

Summary of key learning points:

- real-life scenarios provide the best starting point for good conceptual understanding of multiplication;
- imagery and resources should be used carefully to illustrate features of multiplication;
- as calculations vary, differences in approach should ideally be considered with the intention of selecting the most efficient method;
- instant recall of times tables facts is desirable as this helps to ensure procedural fluency, but conceptual understanding is more important;
- whilst ability to record multiplication using recognised conventions is important, the recording should be purposeful as far as possible;
- all three laws of arithmetic can be used effectively to support multiplication calculations.

Self-test

Question 1

Is 200 an exact multiple of 7? (a) true, (b) false, with a remainder of 4, (c) false, with a remainder of 0·571428

Question 2

The distributive law allows us to split numbers up in order to multiply them. Which of the following would help us to calculate 29×14? (a) $(29 \times 10) + (29 \times 4)$, (b) $(25 \times 14) + (4 \times 14)$, (c) $(30 \times 14) - (1 \times 14)$

Question 3

If milk costs 85p per litre, buying a carton with 0·8 l it should cost you about (a) 74p, (b) 68p, (c) 60p

Question 4

The product of 2, 70 and 800 is (a) 112,000 (b) 11,200 (c) 872

Self-test answers

Q1: (b) 200 is not an exact multiple of 7 as $7 \times 28 = 196$ which is four short of 200, therefore giving us a remainder of 4. Using a calculator to divide 200 by 7 (based on division being the inverse operation to multiplication), you would get an answer of 28·571428, given to six decimal places and noting that this number continues beyond what the calculator display shows us. This is not a remainder as such, but does tell us that 200 cannot be an exact multiple of 7.

Q2: Answers (a), (b) and (c) all work. The first is a traditional tens and units split; the second approach makes use of the fact that multiples of 25 are relatively easy to work out; the last one capitalises on the fact that $29 = 30 - 1$ and recognises that multiplying by 30 will be quite straightforward, before adjusting the answer by subtracting the extra 14.

Q3: (b) 85p × 0·8 = 68p. A repeated addition structure is not helpful here since our multiplier is a number less than 1. Note too that as the focus here is on price, we do not specifically need to know that the 0·8 is in litres.

Q4: (a) The term 'product' implies multiplication and $2 \times 70 \times 800 = 112{,}000$. To give the final answer the numbers have been added, essentially calculating the 'sum'.

Misconceptions

'I have used multilink to work it out and I think $3 \times 2 = 5$'

'I know that when you multiply by 10 you have to add a zero to the end of the number, so 5·2 × 10 = 5·20'

'Multiplication always makes bigger answers'

'$57 \times 34 = 43$ using long multiplication'

'$7 \times 19 = 147$ as $7 \times 20 = 140$ and I have adjusted my answer'

5

Division

About this chapter

Building on Chapters 3 and 4, we now explore the sorts of delights and issues we might face in the teaching of division. These are dealt with under the following headings:

- Interpretations of division
  - Sharing versus grouping
  - Chunking
  - Formal written approaches
  - Interpreting remainders
  - The laws of arithmetic in relation to division
- The terminology of division
  - Dividends, divisors and quotients
  - Factors (and multiples)
  - Prime numbers and prime factors
  - Tests for divisibility
- Working with fractions and decimals
  - Division involving fractions
  - Division of and by decimals

What the teacher and children need to know and understand

Whilst division has a chapter of its own to help us focus on specific elements of its teaching, there are clearly going to be many links back to multiplication since one operation is the inverse of the other. The lengthy discussions about topics such as calculating flexibly made in Chapter 3 also hold true for division.

Interpretations of division

Sharing versus grouping

With division the first thing to note is that there are two main ways in which division can be interpreted, commonly referred to as 'sharing' and 'grouping'. Unless we are consciously incorporating both of these facets into our teaching, we are doing the children a great disservice. Unlike multiplication, where there is a subtle difference between 2×3 and 3×2, when presented with a division calculation it is not possible to tell which interpretation is meant. Consider $10 \div 2$. If no contextual information is given, we have no sense of whether sharing or grouping is being implied and our interpretation could be either: ten broken up into five twos or into two fives. In many ways it does not matter which, as the answer will be 5 either way, but a limited mental concept leaves a learner unable to think flexibly regarding division.

The sharing model is typically the one used in the early stages, with children thinking about how they can share apples or sweets or whatever between a certain number of people.

Common misconception

'12 ÷ 3 = 3 because if I share twelve sweets with my friends we get three each'

This misconception arises because in real life when we share with other people we tend to include ourselves in the share; thus sharing with three friends means that there will be four people sharing altogether. When using a sharing context help children to distinguish between 'sharing *with*' and 'sharing *between*', thinking about how many people the problem implies.

Use of sharing contexts is fine but plan in use of other contexts too, to ensure both facets of division are illustrated. This is exemplified in Figure 5.1 in relation to $10 \div 2 = 5$.

Matthew was given ten books for his birthday. He shared them out into two equal piles, putting one pile on the shelf and packing the other pile to take on his camping holiday (two fives).	
Finley went to the theatre to see a production of 'Noah'. When ten animal characters came onto the stage they paired off into twos (five twos).	

Figure 5.1 Two interpretations of $10 \div 2 = 5$

The relationship between division and odd and even numbers clearly relates to whether a number is divisible by 2, with numbers deemed even either because they divide into two equal groups, or because they can be organised into groups of two. Numicon, a resource we met in Chapter 2, illustrates very clearly the neatness of the two groups and groups of two; whereas the odd-number Numicon pieces always have an extra bit sticking out.

Incidentally, working with non-integer numbers may prompt greater awareness of the relevance of grouping over and above sharing. If all my early experiences of division relate to sharing between people, a calculation like 10 ÷ ½ becomes problematic since half a person is nonsensical. If, however, I have a reasonably well-developed concept of grouping, I can potentially switch to such a scenario and consider how many halves there are in ten (there are twenty), just as I earlier considered how many twos there were in ten.

Common misconception

'Answers in division are always smaller than the starting number'

Since there are twenty halves in ten, this is clearly not the case.

Chunking

'Chunking' is a method which draws on the principle of grouping. It is an informal method ideal for children to make sense of division involving more complicated numbers (large, lots of decimal places, and so on) and this has implications for the calculations we present children with for the purpose of chunking. As with addition, subtraction and multiplication, mental approaches should be considered first, but as calculations get harder, children may need to resort to written approaches. Chunking tends to involve 'jottings' to keep track of the numbers involved and one of the main advantages is that it allows children to retain a sense of the size of the numbers. Essentially division can be considered 'repeated subtraction': the idea that I take my starting amount and take out groups of a certain size over and over again. Chunking builds on this, aiming to gradually take more efficient 'chunks'.

Let us imagine we have a box of 352 chocolates (yummy!) and the calculation 352 ÷ 16. We could contemplate sharing the box between 16 people and seeing how many they get each (sharing), or instead we might imagine filling some smaller boxes (which can each fit 16) and think about how many boxes we can fill (grouping). Approaching calculations from the grouping perspective tends to work well for large numbers, but the learner needs to appreciate that we will get the same answer to the calculation even if the scenario suggests 16 groups rather than groups of 16. Remember: a calculation alone does not tell us which image is the right one.

Using repeated subtraction in its most basic form, we could take 16 at a time from the original 352; we will eventually have all the sweets grouped in sixteens and can then count to see how many groups. Rather than necessarily viewing this as a subtraction process, however, it is entirely possible and perhaps easier to count *up* towards the amount we are trying to divide: 16, 32, 48, 64 and so on. To consider why this might be the case, imagine calculating how many fives in 47 ($47 \div 5$). Using repeated subtraction this would be 47, 42, 37, 32, ..., whereas counting up to work it out we would immediately meet far more familiar numbers, 5, 10, 15, 20, ..., perhaps realising as we count that we already know how many fives in a 'chunk' of the number, maybe that there are eight fives in 40.

The larger the number, however, the longer this would take, and children should be encouraged to think about whether they could take a larger 'chunk' at a time (see, for example, activity 5.10). Perhaps the easiest starting point is to think about whether we have more or fewer than ten of the number in question. In the case of the chocolates we have more than ten, as $16 \times 10 = 160$; some of you will have spotted that double this (16×20) equals 320. There are therefore at least 20 sixteens in 352. Whether we approach this through thinking about the multiples of 10 twice, or go straight for the multiple of 20 (which would be more efficient), we now only have 32 to consider and hopefully our number skills are strong enough to spot that $16 \times 2 = 32$. All the chocolates have now been allocated: 320 (16×20) and 32 (16×2) so $352 \div 16 = 22$ and, returning to our original scenarios, either each of the 16 people will get 22 sweets or we will be able to fill 22 boxes with 16 apiece. Note that the partitioning of the original number into $320 + 32$ splits the number into what might be considered sensible parts, with the more traditional hundreds, tens and units split of $300 + 50 + 2$ being less useful since sixteens do not divide neatly into any of the parts. Helping children to identify the best ways in which to partition numbers is part of our role and is assisted where children are encouraged to share their ideas with each other.

As we have now seen, we can chunk up as well as down, and take chunks of different sizes to arrive at the same answer. This has implications for children's recording of division attempted in this way; informal jottings should be considered personal to the child and will therefore vary in classrooms where children are being encouraged to think for themselves. As a teacher I would want to choose ways of modelling chunking which complement the children's approaches; and where the purpose of recording the mathematics is to explain the *process* to another mathematician (rather than to arrive efficiently at the right answer), children should be encouraged to reconsider jottings which appear unintelligible to others. Sharing the different ways in which people record informal calculations helps us all to gather good ideas from other mathematicians, but there is essentially no right or wrong approach if communication is clear and the answer is accurate and efficiently obtained.

Incidentally, we could have approached this calculation from a sharing perspective, mainly because the number we are dividing by lends itself to this approach as seen in Figure 5.2. Our starting point is $352 \div 2 = 176$, and if we halve again we have divided by 4; the next halving takes us to eighths and a final halving to sixteenths. Had we been dividing by 17 this approach is not one I'd consider!

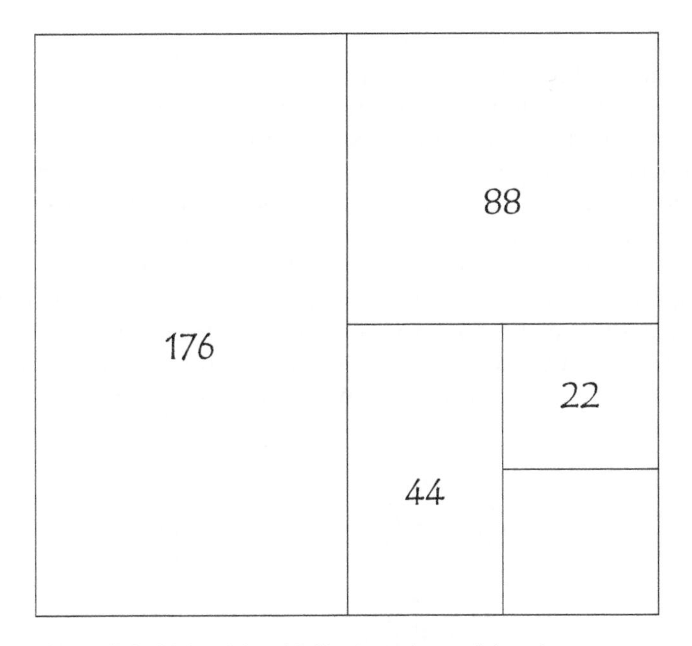

Figure 5.2 352 ÷ 16 = 22 illustrated as a 'share'

Formal written approaches

Short division and long division, both of which are illustrated below, are also based in grouping rather than sharing, but to a certain extent the size of the numbers can be lost since the number is treated as a series of digits rather than as a number as a whole. Such approaches therefore rely on a secure understanding of place value. To calculate 364 ÷ 7 using short division we start by saying 'sevens into three' and then, when we find that there are no sevens in 3, we ask ourselves how many there are in 36. As 7 × 5 = 35 this gives us an answer of 5 (recorded above the top line) and a remainder of 1 (carried over to the next digit). The value of the number we have just used is of course 350 rather than 35, and so in reality our answer is going to be fifty-something. We now consider the final digit, which was 4 but has become 14 because of the 1 carried over from the tens column to the units, and 7 × 2 = 14 so this gives us an answer of 52.

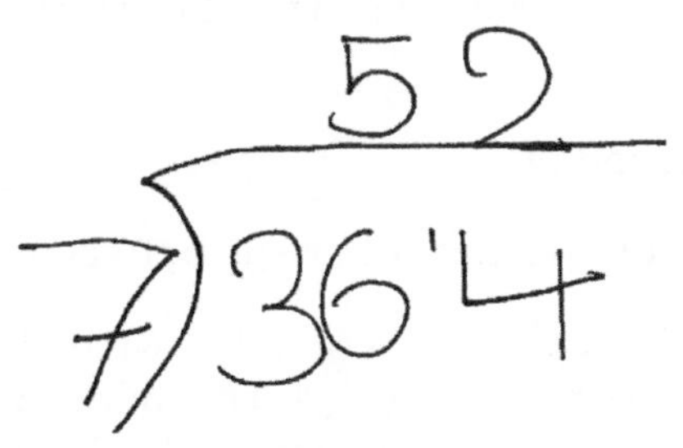

Figure 5.3 Short division

Before you read on, consider how you might have approached this calculation, or that seen in Figure 5.4 if using chunking as your method. Many adults would choose to use a calculator, and actually, in this modern age, this is sensible enough where the numbers involved warrant it. What we would want, however, irrespective of whether we are doing the calculation on paper or fetching a calculator, is to have a rough idea of the size of the expected answer through estimation. Written methods can go awry; I can press the wrong buttons by mistake or forget a decimal point and unless I have some sense that my answer seems to be too big or too small, I may not spot my error. The process is essentially the same for long division as for short division, just laid out a little differently to show the used portions of the starting amount being subtracted as we go. This approach is more useful where particularly large numbers are involved. Learners using any written approach often create a list of multiples at the side as we have in Figure 5.4.

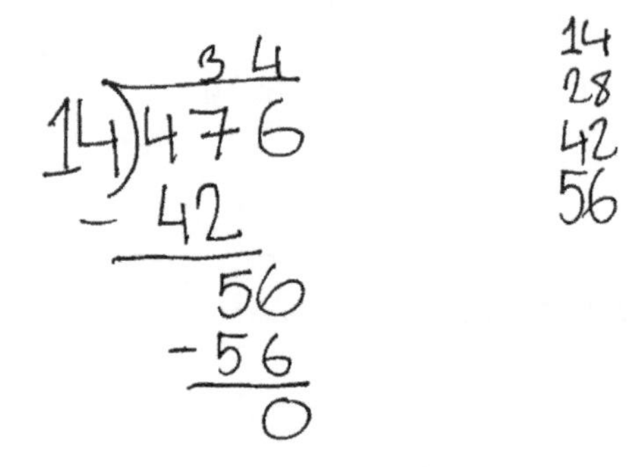

Figure 5.4 Long division for 476 ÷ 14

Interpreting remainders

So far, all the numbers chosen have been nice ones! Calculations without remainders (r) are all very well, but even at a young age children need to experience the fact that amounts do not always share out or group nicely. If there are five cakes and two children, learners need to be aware that whilst a two–three split works in reality (even if it causes arguments), division, like fractions, demands equal parts. Mathematically we have several options, including saying that the children get two cakes each and that there will be one left over, or that they could have $2\frac{1}{2}$ cakes each. In many ways a context such as this one better models the fact that we can have answers involving parts; this is less obvious with division using a resource like Multilink where children do not tend to consider the possibility of half a cube!

Common misconception

'You can't share ten packets of crisps between four people'

Whilst 4 does not divide exactly into 10, the division is clearly possible. These people can share out the packets of crisps. They might get two whole packets plus half a packet each, or we could give an answer of 2 r2 (two each plus two spares).

But there are some situations where parts are not a possibility of course, as this grouping example demonstrates. If there are 31 children in class today, then groups of five won't work out exactly in PE (see activity 5.7), and nor for that matter will any other group size unless we choose to go with ones or thirty-ones! Because we cannot have parts of a person, in this case it would essentially be 31 ÷ 5 = 6 r1 – but we are not going to make the child sit out and miss the lesson, probably instead having five groups of five and one of six. This reminds us of the potential discrepancy between mathematics (where division creates equal groups) and its application in the real world.

The ability to interpret remainders needs particularly careful teaching for the sorts of reasons we have begun to identify above and, depending on context, answers at times get rounded up *or* down. Both 'stories' below relate to 17 ÷ 6, but note that the outcome would be different on account of the different scenarios.

- Charlie's chickens have laid 17 eggs. She collects all the eggs in half dozens (egg boxes with six holes) so that she can transport the eggs without breaking them. How many boxes does she need? [Charlie will need three boxes. The third box will not be full, but she still needs a box so that the eggs do not break in transit.]
- Charlie's chickens have laid 17 eggs. She is going to sell them in boxes of six. How many boxes can she sell? [Charlie only has two full boxes so can only sell two, in spite of having five spare eggs.]

Just as we would want to include both sharing and grouping problems in our teaching, so we should ensure that the children are given opportunities to think about rounding of remainders. Note that the scenario determines which answer is most appropriate and in the case of Charlie's chickens' eggs, whole-number answers were required. But mathematics offers us other solutions suitable for some scenarios. We have already met the idea of stating a remainder (17 ÷ 6 = 2 r5) and begun to think about fractions (the two children with $2\frac{1}{2}$ cakes each), but will now consider decimals as an alternative. Both short and long division lend themselves to continuing beyond the decimal point and in Figure 5.5 we see the remaining 5 eggs being carried over (essentially creating the value of 50 tenths in the next column). As 50 ÷ 6 = 8 r2 we put an 8 above the line and the 2 is carried over. As we would end up calculating sixes into 20 repeatedly, the answer becomes a recurring decimal.

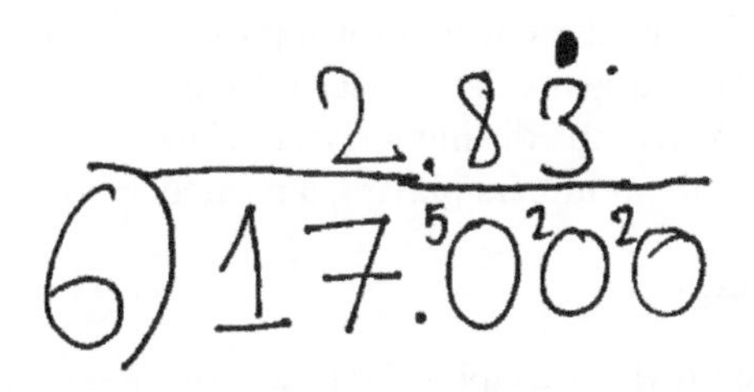

Figure 5.5 Response to 17 ÷ 6 = involving a decimal remainder

If you'd like to test what this particular remainder would be as a fraction, then try question 2 in the self-test section at the end of the chapter!

The laws of arithmetic in relation to division

In the previous chapter we saw that multiplication was commutative; division, however, is not. Try to convince me that $6 \div 2$ is the same as $2 \div 6$ and you will fail! With the associative law we are changing allegiances rather than order but division is not associative either; $(12 \div 6) \div 2 \neq 12 \div (6 \div 2)$ as working out each side gives us answers of 1 and 4 respectively.

Whilst the distributive law does work for division it needs careful handling since only the starting amount can be partitioned. This law allows us to break numbers up, for example, into tens and units, and this means we can deal with large numbers bit by bit, just as we did when chunking which makes good use of this law. For $352 \div 16$ we split the 352 into $320 + 32$, dividing 16 into each part separately. This worked well, however, were we to try to split the 16, into (say) $10 + 6$, this would prove to be unhelpful. Imagine I began by sharing the 352 chocolates between ten of my friends, I would then have none left for the other six; they might not be best pleased!

Common misconception

'$460 \div 12 = 276$ as $460 \div 10 = 46$ and $460 \div 2 = 230$ and $46 + 230 = 276$'

Here we see the distributive law being abused as the user has split the 12, which is inappropriate. Hopefully they went on to realise that the answer was too big to be reasonable.

The terminology of division

Dividends, divisors and quotients

When talking about the calculation $352 \div 16 = 22$ there are certain terms we could have used to describe the numbers. The initial amount to be divided is referred to as the 'dividend', in this case 352. The 16, called the 'divisor', tells us *either* the number of sets/groups *or* the number in each set and of course we cannot tell which unless we have a particular context in mind. Finally, the answer in a division calculation is termed the 'quotient'. Whilst these words form part of our professional lexicon rather than being words we would generally expect primary children to master, we would seek to avoid examples where the distinction lacks clarity, such as $9 \div 3 = 3$, if we want the children to focus on the numbers playing a particular role in the calculation.

Factors (and multiples)

'Multiples', met in the previous chapter, helped us to describe numbers achieved by repetition of an amount; for example, 15 is a multiple of 5 as I can count in fives

(5, 10, 15) and arrive at the number 15. It is also by default a multiple of 3 (3, 6, 9, 12, 15), so three fives and five threes both make 15. 'Factors' effectively take the opposite perspective, describing those numbers which fit neatly inside another, repeated however many times. Rephrasing '15 is a multiple of 3 and a multiple of 5', we can now state '3 and 5 are factors of 15'. Note that these factors come as a pair, something we will test out in a moment using a larger number. The only other pair of factors are 1 and 15 in this case; nothing else divides neatly into 15.

Consider now the number 225 and the process we might go through to identify its factors:

- 1, 225 are unsurprisingly the first pair I have thought of: 1 and the number itself will always be factors.
- Without needing to check, I know that 225 is not divisible by 2 as it is an odd number. This counts out all other even numbers as factors since they will only generate even multiples.
- Appreciating that all numbers ending in 5 (or 0) are divisible by 5 gives me another factor of 225, but what are there five of? Well, 225 ÷ 5 = 45 so this gives me another pair of factors: 5, 45. I must be able to ignore 10 as a factor as 225 does not end in 0.
- I have already discounted all even numbers, so let us think about 7. I know 7 × 3 = 21 so 7 × 30 = 210. This leaves another 15 to think about but as 7 × 2 = 14, no, 7 is not a factor.
- What about 9? Actually 9 must be a factor as it is a factor of 45, one of the other factors we have already found. Its partner is 25 (9 × 25 = 225) so we now have another pair of factors: 9, 25.
- Similarly, the 9 tells me that I must have overlooked 3 as a factor (9 ÷ 3 = 3). We had 9 paired with 25, so having divided 9 by 3 I will need to multiply 25 by 3 to give me the other factor of the pair. Sure enough, 3 × 75 = 225 and so we have found another pair of factors: 3, 75.

So far we have considered 1, 2, 3, 5, 7 and 9, some of which turned out to be factors, and have ignored 4, 6, 8 and 10, just as we can ignore all other even numbers. So just how far do we have to go? Luckily not too far! 1, 3, 5 and 9 all proved to be factors and came with a larger partner, so if we kept going for too long we'd start to find numbers we have already come across and our task would definitely be complete. We still have a few more to check though…

- Might 11 be a factor? 11 × 10 = 110 so 11 × 20 = 220, pretty close to 225 so clearly 11 is not going to be a factor.
- What about 13? I resorted to a calculator for this one. 225 ÷ 13 = 17·307692 (and the calculator display chops off any remaining digits), so clearly 13 cannot be a factor of 225 or it would have given me a nice, neat, whole-number answer.

- The next odd number to check is 15 and probably should have been considered sooner with 3 and 5 already being factors. Sure enough, 15 is a factor, and in fact as $15 \times 15 = 225$ we now know that 225 is a square number. Thus finding paired factors is not *always* the case. Getting to this point also tells us that we need not go any further: we've crossed a kind of mid-point (essentially the square root) and the next factor will be 25, which we have already identified as the partner to 9.

Consider all the various skills employed above to identify the factors of 225; you might want to test your own skills at the end of the chapter!

Factors are not to be confused with number bonds (see Chapter 3). There are lots of pairs of numbers that can be added together to make 225, but for factors our focus is on numbers that divide exactly into 225 or can be multiplied to give it.

Secure factor knowledge is particularly useful in association with fractions which can often be simplified once common factors have been identified. A good mathematician is open to the possibility of common factors, and this includes those beyond the regular times tables; $\frac{26}{65}$ may not appear to be a very user-friendly fraction, but hidden inside is the fraction $\frac{2}{5}$ as $26 = 13 \times 2$ and $65 = 13 \times 5$.

Prime numbers and prime factors

Given the freedom to explore division of different numbers, children will naturally come across examples with numerous factors, some with only a few, and prime numbers: those which have no factors other than 1 and the number itself (see, for example, activity 5.6). Note that 1 is not generally considered a prime number, with prime numbers being greater than 1 and having exactly one positive divisor other than 1.

Occasionally, rather than being required to find the factors of a number, we are asked specifically to find 'prime factors'. The easiest approach is to find any pair of factors you can think of, but then to consider whether those numbers have any further factors; assuming either or both of them do, then you have not yet reached the prime factors and you need to keep going (see Figure 5.6, which uses a factor tree).

Prime factors are particularly useful when we want to compare two different numbers. For example, to identify their highest common factor or lowest common multiple we might start by finding out about the prime factors the numbers share. It is this kind of in-depth knowledge of the number system that allows good mathematicians to manipulate numbers very comfortably.

Producing prime factor trees for 130 and 156 we find:

- $130 = 2 \times 5 \times 13$
- $156 = 2 \times 2 \times 3 \times 13$

The shared factors are a 2 and a 13, and multiplying these gives us the highest common factor: 26 (the largest number that divides into both 130 and 156).

To find the lowest common multiple for these two numbers, we need to take account of the prime factors required by both: 130 needs a 2, a 5 and a 13; 156 requires an extra 2, plus a 3. This gives us $2 \times 2 \times 3 \times 5 \times 13 = 780$ and we can check that

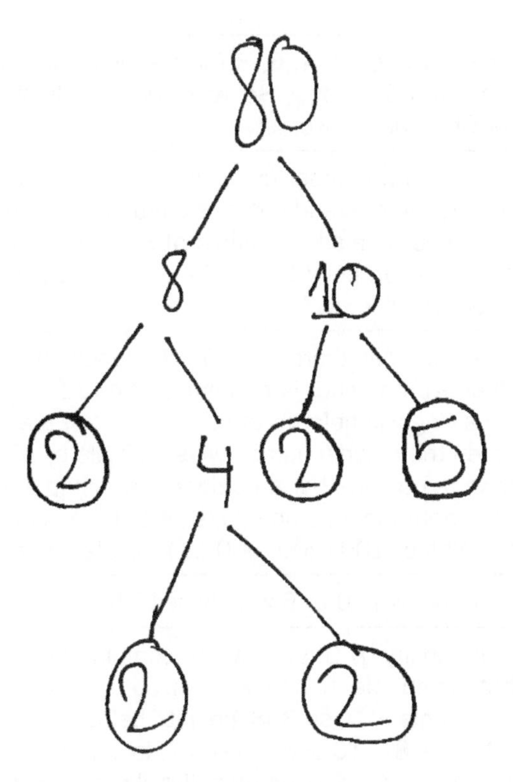

Figure 5.6 Factor trees are a straightforward way of working towards prime factors. Note that the first split need not have been 8 × 10 but could equally have been 2 × 40 or 4 × 20; no doubt you can think of other options

this really is the lowest common multiple by counting in 130s and 156s to check that this is the first multiple we get to which occurs in both patterns:

- 130, 260, 390, 520, 650, 780
- 156, 312, 468, 624, 780

So 780 is definitely the lowest common multiple.

Tests for divisibility

The multiplication patterns discussed in Chapter 4 inform this section and in many ways we have already reinforced the key tests for divisibility in this chapter – for example, recognising that 225 was a multiple of 5 because of how it ended. Let us consider this and the other clues we have as to divisibility, noting that sometimes whilst there is no test, we can at least discount some numbers as factors (see Figure 5.7). Sadly, there are fewer clues for some numbers: the 7× table uses every digit in the units column, can be odd or even, does not add to anything in particular.

÷ 2	If a number ends in 0, 2, 4, 6, 8 (all the units we associate with even numbers), then it divides by 2. So we know 358 will divide by 2 even if we do not know what the answer is!
÷ 3	Adding the digits of a number until you get to a single digit is referred to as finding its digital root, and if this is a multiple of three (3, 6 or 9), then the original number is also a multiple of 3 and will therefore be divisible by 3. A simple case in point is 51; 5 + 1 = 6 and this is a multiple of 3, so 51 must be divisible by 3.
÷ 4	With any even number there's a 50:50 chance that it will divide by 4 as every other even number is a multiple of 4 (32 is, 34 is not, 36 is, 38 is not, ...). Halving will help us to test this out: if the result of halving is also even, then the original number was divisible by 4. The other fact that helps us to work out whether a number is a multiple of 4 is that we only need to think about the tens and units part of the number. As 100 divides by 4, all multiples of 100 (200, 300, ...) will also be divisible by 4.
÷ 5	Every number ending in 0 or 5 will divide by 5.
÷ 6	For testing divisibility by 6, we have to combine two approaches. Firstly, we know that the number has to be even, but secondly that the digital root has to be a multiple of 3, so 358, despite being even, cannot be divisible by 6 since 3 + 5 + 8 = 16 and 1 + 6 = 7. In contrast 3**48** is a multiple of 6; it is even and 3 + 4 + 8 gives us a digital root of 6. If we find a multiple of 6, we have, by default, found a multiple of 3.
÷ 8	Multiples of eight occur every fourth even number so you have a 0·25 chance of any even number you pick being a multiple of 8; and, just like multiples of 4, we can test by halving and in this case halving again. As with multiples of 4, once we get to large numbers we can cheat a little: 1000 divides by 8, therefore all multiples of 1000 are also divisible by 8 and we only need check the hundreds, tens and units portion of a number. This means we know for sure that 39,808 divides by 8.
÷ 9	A digital root test is available to us to check for multiples of 9, with the digits needing to add to 9. So 531 must be a multiple of nine as 5 + 3 + 1 = 9. By default it must also be divisible by 3.
÷ 10	If a number ends in 0 it is divisible by 10 and by default divisible by 5.

Figure 5.7 Common tests for divisibility

Working with fractions and decimals

Division involving fractions

Division involving fractions occurs in three main situations:

- a whole number divided by a fraction;
- a fraction divided by a whole number;
- a fraction divided by another fraction.

Fractions also occur in answers of course, and are particularly pertinent where two whole numbers are being divided and the divisor is larger than the dividend. The very act of writing the division calculation as a fraction effectively turns it into the answer; for example, $3 \div 5 = \frac{3}{5}$ and $6 \div 21 = \frac{6}{21} = \frac{2}{7}$.

At the start of the chapter we met the idea that it was more straightforward to think in terms of grouping when dividing whole numbers by fractions as the act of sharing things out between people was tricky when the divisor is a fraction; half persons are not an easy mental image! Thus $10 \div \frac{1}{2}$ was more usefully interpreted as 'how many halves are there in 10?' But what if it had been the other way round and we wanted to know how many tens there are in a half? Let us explore this in the context of cutting up apples, either through mental imagery or by getting actual apples and a knife if you wish! Just as calculating $3 \div 5$ gave us a fractional answer, so too will $\frac{1}{2} \div 10$ because we are starting with a dividend smaller than our divisor. If we think about this in terms of wanting a group of 10 apples but only having $\frac{1}{2}$ an apple, then we only have $\frac{1}{20}$ of what we need; so $\frac{1}{2} \div 10 = \frac{1}{20}$ or 0·05. Perhaps read on and come back to this in a minute if you need time to think it through.

The calculation $\frac{1}{2} \div \frac{1}{4}$ also makes very little sense if we try to take a sharing approach: how much of the half apple does a quarter of a person get? Remember, altering the emphasis linguistically is helpful, so instead of saying 'a half divided by a quarter' it might be better to opt for 'how many quarters are there in a half?' If I picture half an apple I can visualise the fact that there are two quarters, giving us $\frac{1}{2} \div \frac{1}{4} = 2$. If you are not quite convinced, move back into whole numbers temporarily to go through the same process: $10 \div 2$ means 'how many twos are there in ten?' and as there are five twos in ten, $10 \div 2 = 5$. Making both the denominators the same can also help if you are still not sure, so this question becomes $\frac{2}{4} \div \frac{1}{4}$ and I can now effectively ignore the denominator and just focus on the division of the numerators: $2 \div 1 = 2$. Being allowed to ignore something in mathematics can feel terribly wrong, but it works in this case as long as the denominators are identical. I am sorting a number of quarters out into single quarters and I want to find out how many there are. It could just as easily be *any* size slices of apple; if I'm sorting out $\frac{8}{15}$ into groups of $\frac{2}{15}$ (though goodness knows why), there will be four groups as $8 \div 2 = 4$. Had I had $\frac{14}{15}$ of the apple to start with, then I'd end up with seven groups.

To tackle $\frac{1}{4} \div \frac{1}{2}$ we will go through a similar process to $\frac{1}{2} \div \frac{1}{4}$, but this time we will not be expecting a whole-number answer; there are therefore some similarities to $\frac{1}{2} \div 10$. Picture or cut yourself quarter of an apple; asking how many halves you have got may seem rather a daft question, but the key here is to realise that you have only part

12 ÷ 0·3 = 40	0·3 is smaller than 3, therefore we are expecting a bigger answer since more of them will fit into 12. The divisor being ten times smaller results in a quotient ten times bigger.
12 ÷ 0·03 = 400	0·03 is smaller still, a hundred times smaller than 3, so the quotient this time will be a hundred times bigger.
1·2 ÷ 0·3 = 4	Both the dividend and the divisor are ten times smaller than the original calculation and an estimated answer will confirm that the answer of 4 must be right as 0·4 would be too small and 40 too big.
1·2 ÷ 0·03 = 40	The arguments may be beginning to sound familiar; that 0·03 is smaller and will therefore fit more times into 1·2.
1·2 ÷ 3 = 0·4	This is the first example where the divisor is larger than the dividend. This has entirely the same effect as our earlier fraction examples: the answer must be less than one.
0·12 ÷ 0·03 = 4	This is another example which is equivalent to the original calculation, with the dividend and divisor both a hundred times smaller.
0·12 ÷ 0·3 = 0·4	Here I'd probably multiply both numbers until I have either 1·2 ÷ 3 or 12 ÷ 30 as these somehow seem easier to get my head round, and I have a better sense of the size of number I am expecting.

Figure 5.8 Mental calculation involving division of decimals

of what you wanted – only half a half! So $\frac{1}{4} \div \frac{1}{2} = \frac{1}{2}$; if you want to check your grasp of this, have a look at question 4 in the self-test section.

Division of and by decimals

There are several possible approaches to consider when faced with division involving decimals, and as with all other types of calculation, mental methods ought to be considered first, especially where the numbers involved are variations on known facts and therefore not too complicated. The table in Figure 5.8 illustrates this by exploring variations on 12 ÷ 3 = 4, focusing on grouping as opposed to sharing as this again makes the concepts and relationships easier to grasp. We have 12 ÷ 3 = 4 which tells us that there are four threes in 12, and this forms the basis of all the other calculations.

Did any of the answers in Figure 5.8 surprise you? The third one often catches people unawares if they are not anticipating any answers the same as the basic 12 ÷ 3 = 4, but because both the dividend and the divisor are ten times smaller, this has no effect on the answer; similarly the sixth one where both were a hundred times smaller. The calculation 120 ÷ 30 would also equal 4, as would 1200 ÷ 300. Another way of looking at this and other calculations is to consider the numbers in a fraction format, and to switch to an equivalent fraction, for example, by multiplying top and bottom by 10 in the case of $\frac{1.2}{0.3} = \frac{12}{3} = 4$. This is a particularly helpful method since it gets rid of the decimals and means the calculation can be solved using whole numbers instead. Note that as long as the same ratio is maintained, the changes do not have to be in multiples of 10. Doubling numerator and denominator, $\frac{3.5}{28}$ becomes $\frac{7}{56} = \frac{1}{8}$ or 0·125.

Seeing numbers written as a fraction may prompt you to consider cancelling, and this can negate the need to resort to a more formal written method. For example, for 6·3 ÷ 1·05 we might go through the steps as follows: $\frac{6.3}{1.05} = \frac{630}{105} = \frac{126}{21} = \frac{42}{7} = 6$. Hopefully you can see what I have done at each stage to arrive at the answer: 6·3 ÷ 1·05 = 6. Had we wanted or needed to employ a formal written method we would ideally have made the divisor a whole number as we did here, and as I would choose to do for 27·72 ÷ 1·1, effectively turning the calculation into 277·2 ÷ 11. This is not an example I would want to attempt to cancel, using short division instead as illustrated in Figure 5.9.

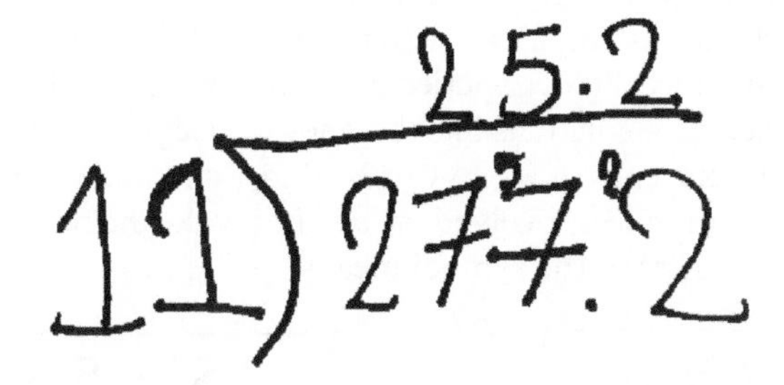

Figure 5.9 Division of a decimal using a formal written approach

This seems a reasonable answer since 25 × 10 = 250 and the numbers involved here are slightly larger, so we can be fairly sure that we have not misplaced a decimal point. But when dividing with decimals, if all else fails, grab a calculator!

Where remainders are generated, use of the ratio approach needs to be somewhat more cautious. If 560 ÷ 24 is about how many packets of biscuits I can make from a starting number of 560, putting 24 in a packet, my answer is 23 r8; in other words, 23 complete packets and 8 biscuits left over. But were I to present $\frac{560}{24}$ as a fraction and cancel this by dividing top and bottom by 4, my resulting calculation, 140 ÷ 6, would suggest a remainder of 2. This is a quarter of the size it should be and happens because we divided by 4 along the way. Whilst not an insurmountable problem, it clearly requires awareness and careful handling!

Activities for the teaching of division

Consider the following activities as a mix and match opportunity; whilst designed to target a particular area of learning, most of them can be altered to switch the focus in some way. Remember that some teaching need not focus on answers but on raising children's levels of awareness instead. Earlier in the chapter we established that 39,808 must be divisible by 8; we may not have any idea how many eights, but we know it will be a whole-number answer. Build in context where you feel it will support the children's understanding and remember to emphasise 'grouping' as well as 'sharing' in your examples.

Pupil activity 5.1

Familiarity with the symbol for division
Y1 to Y2

Learning objective: To appreciate how division calculations are recorded and interpret '÷' correctly in both contexts.

The division symbol tends to be the last of the operations to be introduced, and this activity builds on the children's familiarity with the others. It is assumed, however, that the children already have a wealth of experience of division in practical terms. Present a small range of completed calculations like the ones we see here and ask the children to work in pairs to 'translate' each one into words. Either children will suggest a meaning for '÷' because of the numbers involved or it will provoke the necessary discussion to allow the introduction of what the symbol means.

$10 - 2 = 8$	$15 = 5 \times 3$
$11 = 4 + 7$	$12 \div 2 = 6$

Later use of the division symbol should include examples where the divisor is either 1 ($14 \div 1 = 14$) or the same as the dividend ($14 \div 14 = 1$), since we know such examples can cause confusion.

Pupil activity 5.2

Families of division facts
Y2 to Y6

Learning objective: To recognise that one division fact leads to others.

Initially children will be working practically in division, hopefully experiencing both sharing and grouping contexts. Once they start recording division using formal symbols this is an ideal time to start drawing children's attention to related facts, and this activity focuses on the links between the calculation and suitable pictures. For example, $8 \div 4 = 2$ and $8 \div 2 = 4$ are closely related since either of them could be used to represent a calculation which results in two fours or four twos, both of which are illustrated in this picture. Links with multiplication can also be made clear, with 4 × 2 = 8 and 2 × 4 = 8 completing the set. Children should match individual calculations with pictures showing both versions together, noting they should expect to find more than one calculation for each image. See activities 4.6 and 3.3 for further ideas on recognising relationships between facts.

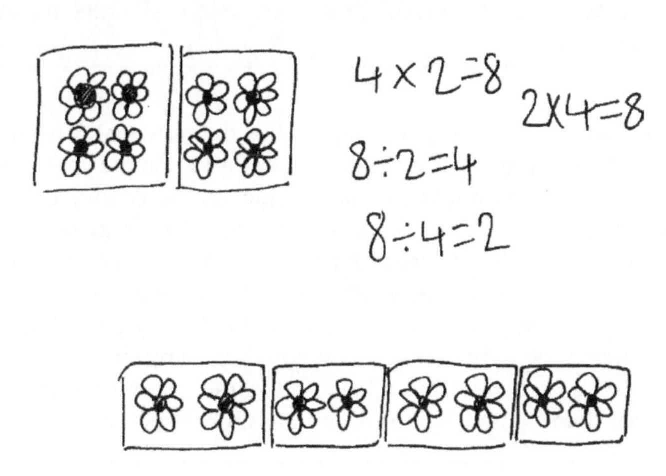

For older children more complex sets of calculations without pictures would be more appropriate, identifying the relationships between calculations with the same numbers but where some involve decimals and others are multiples of 10. This can be done collaboratively by putting the answers up around the room; given a particular calculation, children should turn and point to the answer they think is the right one. For example, which answers would the following calculations require?:

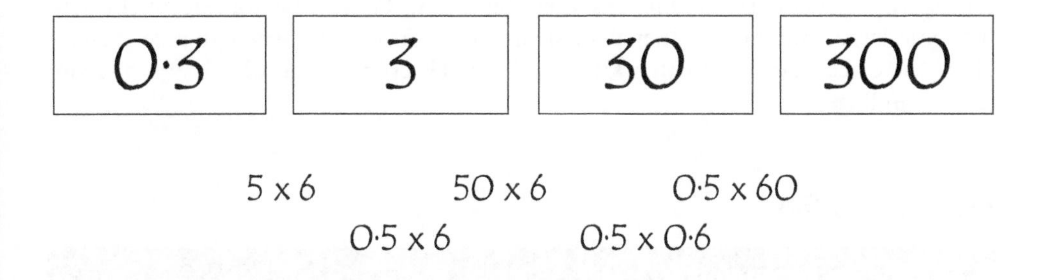

You could of course throw in some red herrings: what would you want to point to for 50 × 60?

With enough experience, older children can begin to make more advanced links. For example, rather than using a written method or chunking approach to solve 81 ÷ 3, we might use our knowledge that 9 × 9 = 81. Since a nine consists of three threes, there must be 27 threes in 81.

Pupil activity 5.3

Investigating odds and evens
Y1 to Y3

Learning objective: To understand that the units digit tells us whether a number is odd or even.

Appreciating division as sharing and grouping has the potential to support children in understanding that a number is even either because it can be split neatly into two equal groups or into twos. Numicon provides an ideal resource for illustrating this in a really visual way and could be used as part of this activity to emphasise the two possibilities. In particular, draw the children's attention to the fact that 10 is even since this underpins children's ability to appreciate that a larger number's status as odd or even depends only on its unit digit since the tens part will definitely divide by 2. Other base 10 equipment can be used to the same effect; however many sticks of 10 I have, every one of them could be broken neatly in half, and in this respect the emphasis on two equal parts (rather than groups of two) is perhaps more useful as numbers get larger.

Common misconception

'47 starts with an even number so it must be even'

Statements such as this are common where children do not fully understand the concept of odd and even numbers.

Children should pick from a range of two-digit numbers and specify whether each one is odd or even, articulating why they think this. Those who want to might then try some larger numbers; rather than shying away from extremes, embrace them; a child who can articulate how they know 1,000,000 is even probably has a pretty secure understanding!

Pupil activity 5.4

Sharing and grouping
Y2 to Y6

Learning objective: To understand that a single division calculation can be interpreted in different ways.

Having noted the importance of exemplifying both 'sharing' and 'grouping' in our teaching of division, this activity encourages children to consciously identify the possible interpretation of division calculations by linking them to real and imaginary scenarios. Prepare cards with calculations on them and specify wanting two different stories (either pictorial or in words, which could be spoken rather than written). The contexts suggested by the pictures will help some children if they are struggling to think of one of their own!

... groups of...		... shared between...
	27 ÷ 3 	
	39 ÷ 13 	
	8 ÷ 2 	

In the long term the children could work just from calculations, interpreting each one in at least two different ways. Scrutinise the children's stories as to whether they definitely are interpreting the calculations as both 'sharing' and 'grouping' as you will find some children's stories appear different on the surface but are actually both examples of the same structure:

- I shared 18 raspberries out and put 2 on top of each pudding.
- I organised 18 shoes in pairs. There were 9 pairs.

Pupil activity 5.5

Sharing numbers
Y1 to Y3

Learning objective: To understand that not all numbers share exactly and begin to suggest solutions where this happens.

Children's early experiences of division are sometimes so tightly orchestrated by the teacher that the only numbers the children meet are those which divide exactly. Where this is the case, we run the risk of children assuming that all numbers divide in this way. What we might do more profitably is to explore various problems, aiming to compare them and identify those which share out easily (neatly into equal groups) and those which do not. With some children we may need to stress that division is about creating equal groups.

I would start with a context where the items will remain whole; perhaps a bag of boiled sweets. Ask the children to shut their eyes and imagine they have so many in their hand to be shared between two or three children. Do they think the sharing will work out fairly? Model and test some together using the sweets, ensuring that some work and some do not. Then give the children the freedom to investigate numbers and shares of their own choosing.

This activity can then be extended to include items, such as cookies, which could be broken to allow the sharing to be completed differently. Pat Hutchins's story *The Doorbell Rang* (1986) would complement this really nicely as more and more children arrive to share the cookies as the story progresses. I will not spoil the ending!

Pupil activity 5.6

Factors
Y1 to Y6

Learning objective: To understand that factors are numbers which divide exactly into another; to appreciate there can be different numbers of factors irrespective of the size of the original number.

A key aspect of factors work is to engage children in making lots of numbers, for example, using Multilink cubes, and then investigating how those numbers can be broken up. The emphasis, remember, is not on number bonds, but on dividing the original number into equal size chunks. Plan in advance some good numbers for the children to explore, checking that you have some with lots of factors (multiples of 12 are a good choice); some with very few (15 is more limited than 12); plus at least one square number and a prime number for good measure.

Later work on factors will not, ideally, involve physical resources, aiming instead to develop knowledge about possible factors and the skills necessary to check whether they are. Before the children explore some numbers of their own choice I tend to model a systematic approach to this much like we saw earlier in the chapter when seeking the factors of 225.

Pupil activity 5.7

Division in PE
Y1 to Y4

Learning objective: To appreciate that some numbers group exactly, whereas others do not.

PE lessons or similar offer a really good opportunity to explore groupings. I have the children run around first and then on a certain signal they stop and I call out a number which tells them the group size. Once they've gathered the right number of people they sit down and anyone who's left comes to me; a brief discussion about the number of groups and remaining children follows, and then we play a few more times to explore different groupings. We might even try 'ones' as it's important to understand $\div 1$. Aim to finish with group sizes that suit the rest of your PE lesson!

Clearly this can be linked to multiplication and explored through something like arrays where the children can sometimes be organised neatly into rows and columns ($7 \times 4 = 28$, say) and at other times cannot (29 children in PE that day). This is illustrated really nicely in Elinor J. Pinczes' book *Remainder of One* (1995), where Joe the bug uses mathematics to solve his problem!

Pupil activity 5.8

Prime factors
Y5 to Y6

Learning objective: To understand how factor trees can be used to find prime factors.

Published worksheets often constrain the children, forcing them to devise factor trees following a set pattern, so instead give the children a whole range of numbers to explore. At some point ask the children to compare their 'trees' with those of others. This will not only highlight the fact that the trees can look different, but is also a good way of checking that the answers are right, since even a differently structured tree should result in identical prime factors.

When selecting your range of numbers ensure that they are suitably challenging (the prime factors of 15 are just too obvious to warrant a written approach) and that the trees will be varied, with some reasonably extensive (try 160) and others very small (try 221, which is potentially far more challenging to crack). With opportunities to work in small teams, children can be challenged to do things like find all the possible 'trees' for a particular number.

Pupil activity 5.9

The relationship between place value and division
Y2 to Y5

Learning objective: To draw on understanding of tens and units to develop understanding of division.

In many ways the use of place value imagery and apparatus is beneficial because it supports children's appreciation of the sizes of the numbers involved in division. I helped with a Y5 class once who were struggling with calculations like 72 ÷ 6, partly because their sense of how big 72 was needed refreshing. Retrieving base 10 equipment made a big difference and through its use the children began to use chunking more efficiently, soon working without the need for resources. In the case of 72 ÷ 6, partitioning the tens and units into 30 + 30 + 12 is particularly helpful in identifying how many sixes. Think about how the children might tackle:

- 91 ÷ 7 =
- 128 ÷ 8 =
- 76 ÷ 4 =

Incidentally, the last of these would merit discussion since knowing that 80 ÷ 4 = 20 might lead us to the answer rather efficiently, saving the need for resources.

Pupil activity 5.10

Chunking
Y3 to Y6

Learning objective: To appreciate what makes efficient chunking.

Mathematical skills improve with practice, and chunking is no exception. Having lots of opportunities to explore division by grouping, children will begin to realise that several groups, a 'chunk' of the original, can be considered in one go. The aim of this activity is to refine the chunking process, and the calculations have been carefully chosen to be large enough to warrant chunking and to prompt the possibility of efficient chunking by working from known chunks. Note that a calculation such as 21 ÷ 3 would not be suitable, being ideally a known fact. The problems have been set in context as this helps some children to get their head round what is being asked.

- Each fence panel needed 14 nails and Joe had a pack of 200. He needed to make 13 panels; did he have enough nails? [As 140 nails will be enough for 10 panels he definitely has enough nails left for the remaining three as 14 × 3 will be less than 60.]

- The mechanic responsible for assembling the cars' engines knew 32 bolts would be needed per car and had a pack of 800. How many engines could be assembled? [Here we might recognise that 32 × 20 = 640, leaving us to think about how many more engines can be made using the remaining 160. Five more since 160 is half of 320 (32 × 10).]
- The boot manufacturer needed 12 eyelets per boot, so 24 per pair. Bulk packs contained 460 eyelets, so how many pairs of boots could be made? [Had there been 480 in a pack, this would have made 20 pairs; 460 is 20 eyelets short so will only be enough to make 19 pairs of boots.]

Chunking is an informal procedure therefore children should feel able to jot their workings in different ways, but talking about the workings together is vital since it is this that leads children to work more efficiently as they learn from each other. Note that we do not always need actual numerical answers to appreciate efficient chunking: I know that Joe in the first question has enough nails but do not know or need to know how many will be left over. Clearly younger or older children might require different numbers and contexts, but the principles are the same.

Pupil activity 5.11

What method?
Y4 to Y6

Learning objective: Developing flexible approaches to division calculation.

One way of encouraging children to think flexibly about division is to spend lots of time on just a small number of carefully chosen calculations. Thus, rather than generating lots of different answers, the emphasis is switched to generating lots of different methods. Consider the calculation 294 ÷ 3 and the different approaches the children might use, such as chunking or short division. This child chose to combine chunking with a number line approach:

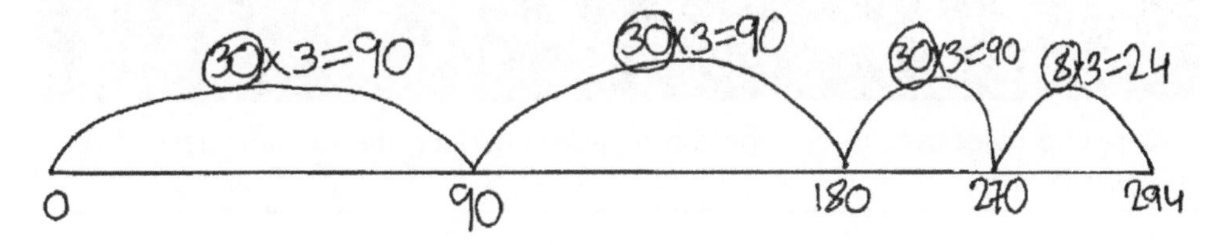

Whilst the children might work individually to start with, collaboration should be built into the lesson to allow them to share their ideas. Which are their favourites? Is it because they are easy, or quick, or something else?

Having an idea of particular methods you would like to promote is important if you want to make the most of your mathematics lessons. For example, this calculation was chosen to see if the canny amongst you had spotted that it's just six short of 300, so as $300 = 3 \times 100$, 294 is a couple of threes less; in other words the answer must be 98, without any need for a complex written method. If you did not spot this, there is no need to panic; trying calculations out for yourself at the planning stage will gradually alert you to such possibilities, and the children themselves will add to them given the opportunity to do so. Trying things out in advance is the easiest way to ensure you are well prepared for the lesson, but will also give you the opportunity to consider whether the numbers involved lend themselves to being worked out in different ways.

Pupil activity 5.12

Formal or informal division
Y5 to Y6

Learning objective: To make appropriate decisions about calculations which warrant written methods.

Rather like some of the activities in Chapters 3 and 4, this one encourages children to study calculations to decide which ones really warrant a formal written method. I would print a ready-made worksheet; children will typically find that some of the answers can be obtained mentally. To test this out, give the sheets out one between two but nothing to write with; can the pairs provide any answers? Share a few ideas as to which are the easiest calculations and why. The children can then use a written method of their choice to answer the remaining questions, though you may want to specify that particular methods are used at least a certain number of times.

Other ways of generating calculations which may or may not warrant the use of formal written methods include picking from raffle tickets, rolling dice or using spinners.

Pupil activity 5.13

Estimation of answers in decimal division
Y4 to Y6

Learning objective: To develop the habit of checking the reasonableness of answers.

Clearly children should be encouraged to attend to the reasonableness of their own answers to division calculations, but one of the easiest ways to force the issue is to give children work with answers, some of which are right and some wrong (though all the answers below are wrong) so that children can be asked to give the right answers.

- 36 ÷ 100 = 0·036 [Ability to spot answers which must be too big, or in this case too small, is crucial and children will benefit from lots of practice. Ability to divide successfully by multiples of 10 is a particularly useful skill.]
- 2 ÷ 4 = 2 [Underpinning a calculation such as this is the essential knowledge that the number system can be subdivided. That there is something inside 2: four halves in this case.]
- £55 shared between four people will be £13·50 each. [Money provides a nice context if children are struggling with the concept of subdividing whole numbers as most will recognise more readily that pounds can be split into pence. Here each person has lost 25p somewhere.]
- A block of cheese weighing 1·25 kg was cut into five equal portions. Were these 25g each? [The answer perhaps recognises that the portions might be described in grams rather than kilograms, but should be 250 g. The loss of the zero could be down to a number of reasons such as not being clear that there are 1000 g in 1 kg.]

Concept cartoons also offer a really nice way of exploring this topic and here we see more than one correct answer:

Pupil activity 5.14

Division involving fractions
Y5 to Y6

Learning objective: To develop understanding of division involving fractions.

As long as the children are working in context they may well surprise you with their ability to work with division of fractions but the fact that things can be subdivided is an essential foundation to this. The apple helps to remind us that this is possible. Here we combine multiplication with division since this enables us to probe the potential misconceptions more easily. Once the children have had a go at making up some calculations, using these numbers in any order, display the calculations anonymously. Children circulate the room deciding which ones they agree with and which calculations they are not sure are right. As ever, discussion will be key in supporting children's developing mathematical understanding.

Pupil activity 5.15

Dealing with remainders
Y1 to Y6

Learning objective: Ability to identify whether remainders should be rounded up or down depending on context.

The success of activities relating to remainders often relates to the teacher's choice of context, ideally choosing those accessible to the children; even very young children can understand the concept of dealing with what is left over from sharing or grouping. The earlier examples involving Charlie's eggs were ideal as both rounding up and down could be illustrated and any activities could be supported by resources, with actual egg boxes a cheap resource, easy to come by. Other contexts which work well to illustrate rounding up and/or down include:

- Numbers of buses. If we need to transport 201 people on a school trip and buses seat 50 people, we will need five buses. [Rounded up]
- Baking. If buns are to be sold in bakers' dozens (13) and we make 300, we will be able to sell 23 packets and have 1 bun left over. [Rounded down]
- Tablets. A patient needs to take two tablets per day and tablets come in packets of 20. If he is going away for 4 weeks, he will need to take three packets. [This

scenario is not as straightforward a division example but is essentially rounded up since the patient will be taking enough tablets for 30 days but will only be away for 28.]

In discussion between yourself and the children you will want to highlight mathematical features and realities: whilst it is really annoying that we need a fifth bus for just one person, we cannot take a chance on someone being away on the day of the trip and just book four buses. Note that with each of these scenarios the answers are whole numbers; unlike the cookies in activity 5.5, we are not contemplating breaking things in half here!

What other contexts would work well for your class?

Pupil activity 5.16

Investigating remainders
Y5 to Y6

Learning objective: Familiarity with when and why we get remainders in division.

In contrast to activity 5.15, this activity is not set in context, rather it gives children a more abstract task: the opportunity to investigate division calculations on the hunt for particular remainders. Looking at the choice of numbers given here, try a few out just to get a feel for it, including finding examples which divide exactly so do not generate a remainder. Did you find any pairs that gave you a remainder of 3? What about remainders of 1? Are there any pairings which you predict will reliably give you certain remainders? How do you know? Encourage the children to explain! Looking for a particular remainder, how will you know you have found all the possibilities?

Start with...	Divide by...
41, 85, 72, 33, 23, 64	7, 5, 4, 2, 3

The extensive questioning and opportunities for discussion are vital if children are going to get the most out of such an investigation, helping the children to appreciate the general situations which give rise to remainders. Open-ended investigations like this are also a great leveller; children who may not usually flourish might surprise you with their insights.

Pupil activity 5.17

Exploring division using a calculator
Y5 to Y6

Learning objective: To develop understanding of the sorts of answers that are possible in division.

Calculators provide a really good tool for exploring the sorts of answers you get when you divide different pairs of numbers; just choosing pairs from the digits 1 to 9 will give us whole-number answers, finite decimals and recurring decimals, with one of each completed below. The two-way table provides a way of recording findings from the investigation. This makes it easier to spot answers which occur more than once and think about why, and to begin to get some sense of where particular answers such as recurring decimals occur. At primary level we do not expect children to understand some of these features fully, but being aware that they exist sows seeds to be germinated at a later stage.

	1÷	2÷	3÷	4÷	5÷	6÷	7÷	8÷	9÷
1				4					
2									
3									
4									
5							1·4		
6									
7									
8									
9		0·222...							

Another way of exploring the different answers we get to division calculations is to draw on measures topics. With real-life measurement scenarios we do not tend to get 'nice' answers; for example, if a box of stock cubes states an overall mass of 71 g, roughly what does one of the 12 cubes weigh? The calculator gives an answer of 5·9166666, so children need to be able to interpret this: each cube must weigh almost 6 g.

Summary of key learning points:

- modelling of early division should include both sharing and grouping contexts, leading to the fact that both answers will be the same;
- real-life contexts will support children in making sense of division;
- division may not lead to whole-number answers, and the remainders can be dealt with in different ways, considering the original context of the question where appropriate;
- building on known division (and multiplication) facts allows us to work out new facts;
- methods for tackling division calculations vary and should be considered flexible.

Self-test

Question 1

210, which is not a square number, has how many factors in total? (a) 15, (b) 16, (c) 14

Question 2

Thinking back to Charlie and her 17 eggs in boxes of six (17 ÷ 6), the remainder given as a fraction would be (a) $\frac{5}{6}$, (b) $\frac{5}{10}$, (c) $\frac{6}{10}$

Question 3

Is the number 123,456 divisible by (a) 3, (b) 4, (c) both 3 and 4?

Question 4

Picturing a quarter of a nice big sponge cake, $\frac{1}{4} \div \frac{3}{4} =$ (a) 3, (b) $\frac{1}{4}$, (c) $\frac{1}{3}$

Self-test answers

Q1: (b) We know that (a) cannot be right; factors come in pairs unless the number in question is a square number. If you thought the answer was (c) it may be that you are missing one of the following pairs: 1 × 210, 2 × 105, 3 × 70, 5 × 42, 6 × 35, 7 × 30, 10 × 21, 14 × 15.

Q2: (a) $\frac{5}{6}$ represents the fact that there were five remaining eggs, a five-sixths part of a full box which would contain six eggs.

Q3: (c) The digital root of 123,456 is 3 (the digits first add to 21), therefore 123,456 must be a multiple of 3. Turning our attention to the tens and units end of the number we see 56 and as this divides by 4 ($56 \div 4 = 14$), then the whole number will also divide by 4. We do not know how many threes or fours, but we know that it will be an exact answer.

Q4: (c) We know the first answer cannot be right, since the divisor is larger than the dividend so it cannot be a whole-number answer. As the divisor (the specified group size) is $\frac{3}{4}$, we need to ask ourselves how much of $\frac{3}{4}$ of a cake we have: as we only have $\frac{1}{4}$, we only have a slice $\frac{1}{3}$ the size of the piece we wanted! Remember that once the denominators are the same (and they already are in this case), we can effectively ignore them, and focus on the division of the numerators, in this case $1 \div 3 = \frac{1}{3}$.

Misconceptions

'12 ÷ 3 = 3 because if I share twelve sweets with my friends we get three each'

'Answers in division are always smaller than the starting number'

'You can't share ten packets of crisps between four people'

'460 ÷ 12 = 276 as 460 ÷ 10 = 46 and 460 ÷ 2 = 230 and 46 + 230 = 276'

'47 starts with an even number so it must be even'

6

Algebra

About this chapter

The term 'algebra' is often more closely associated with secondary education, but, as we will find in this chapter, there is much that can be done at primary level; primary practitioners who appreciate the key features of algebra are able to lay suitable foundations. We explore two main areas, algebra in relation to pattern, and describing unknowns using algebra, noting that the two overlap to a great extent. The two topics are brought together in the final section, where problem solving provides an opportunity to build on the relationships.

- Pattern: a key feature of algebra
  - From 'pretty' to mathematical patterns
  - Number patterns and algebra
- Algebra and its ability to describe unknowns
  - Numbers and variables
  - Notation of algebra
- Algebra through problem solving and investigation

What the teacher and children need to know and understand

Pattern: a key feature of algebra

From 'pretty' to mathematical patterns

Human beings seem to have a natural disposition towards spotting and creating patterns. But as 'pattern' is an everyday word as well as one which is used in the context of mathematics, learners need to appreciate that rather than 'pretty' patterns, perhaps with rather random use of colour, mathematical patterns are based on some sort of repetition, or on a defined sequence of some sort. We will meet the idea of pattern again in Chapter 10 when we think about tessellation and tiling; those patterns are two-dimensional as they cover a surface area. Here our focus will mainly be on patterns which extend one thing after another. Let us think about that in terms of a string of beads like a necklace:

- The easiest possible example, and one we might not consider as a pattern, is to string together beads of identical colour, shape and size (pearls, say).
- The next level of difficulty would be to change just one feature, typically colour, keeping things like size and shape the same. The simplest example would be to alternate between just two colours, stringing one bead of each. A bead pattern that goes 'black, white, black, white,...' repeats after two beads, whereas our pearl necklace had a unit of repeat of only 1.
- Beginning to combine different types of bead progresses the level of difficulty; I might still be using the black and white beads but this time use a mixture of large and small beads as in Figure 6.1. Note that the unit of repeat here is actually 6, despite using only four different types of bead; it is the seventh bead that starts the repeat of the pattern.

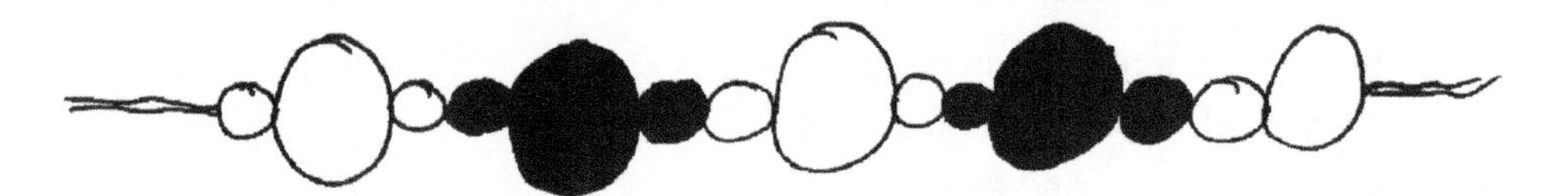

Figure 6.1 Bead pattern with a unit of repeat of 6

- So far the necklaces have involved beads of different colours and sizes; what they could also include would be beads that are different in shape as well as varying in colour and size.

Whilst the above list helps to draw our attention to some different types of pattern, given freedom to explore with resources which lend themselves to generating such patterns the children will most likely create examples of many different types. In addition to the suggestions above, the children might create examples where orientations change (for example, equilateral triangles in different rotations as in Figure 6.2) or patterns in which there is an element of symmetry as seen in Figure 6.3. Sharing patterning choices explicitly with the rest of the class will help everyone to become more conscious of the 'rules' by which the patterns have been made, or the repetitions involved. The foundations are laid for generalisation at this point through ability to articulate how simple patterns work. Necklaces may be less appealing to some children, but visual/physical patterns can be generated in lots of different ways – anything from towers built of Multilink to traffic jams with toy cars.

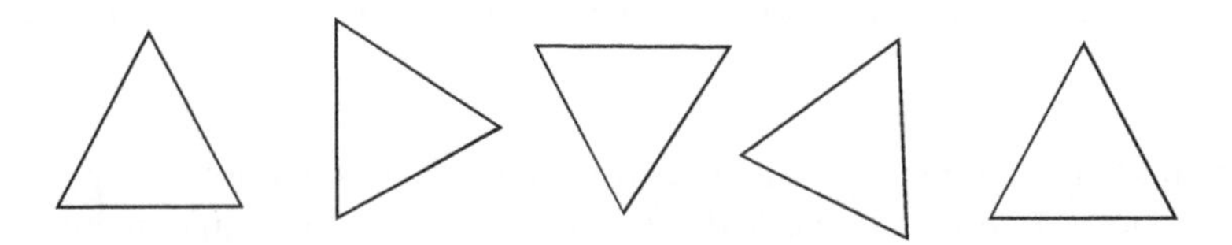

Figure 6.2 Pattern created by different orientations of the same shape

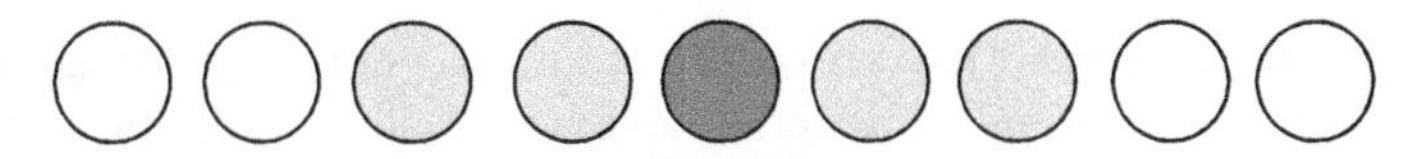

Figure 6.3 Pattern incorporating symmetry

Musical rhythms also lend themselves to exploration and reinforcement of repeating patterns, and are particularly powerful as they draw on auditory rather than visual discrimination skills. A child who is struggling to identify the repetitions in a bead pattern may hear and 'feel' the repetitions, for example, in 'clap, stamp, clap-clap, stamp / clap, stamp, clap-clap, stamp'. For this reason it is helpful to get the children to say their patterns out loud from time to time: 'red, red, blue, red, red, blue, red, red, blue'. Errors made are often picked up by the children through hearing that the pattern sounds wrong. Pattern also manifests itself in everyday scenarios, such as the days of the week, cycling through the same seven days in the same order repeatedly and recognising that 'Sunday, Monday, Tuesday' sounds normal whereas 'Monday, Sunday' does not.

As we have already seen, units of repeat feature in our understanding of pattern, and appreciation of them is fairly essential in enabling us to begin to think algebraically. Whereas a young child may only be able to tell you the colour or shape of the next bead, a more developed sense of the pattern allows us to predict items much further ahead. If the pattern we see in Figure 6.4 continues in the expected fashion, what would the 12th symbol be? That is easy as it is the next one, but what would the 100th symbol be? The 101st? The 99th? Assuming you did not draw the missing symbols in the sequence, which would be a long and boring task, you probably appreciated the unit of repeat and used your knowledge of multiples to work out what symbol would appear in the 100th position. This could have been multiples of 5, the smallest unit of repeat in this case, or perhaps you thought in terms of a double unit of repeat, and imagined the block of ten, repeated ten times until you reached picture 100. However you did it, algebraic thinking was involved.

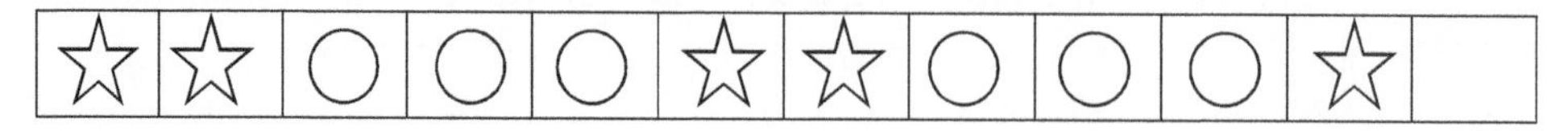

Figure 6.4 The beginning of a pattern: how far ahead can you predict what the symbols will be?

For children to appreciate the way in which patterns continue as you have just done, they need to ask and be asked these sorts of questions. Enjoyment of pattern making is a good start but is not enough on its own.

Number patterns and algebra

So far the patterns discussed have revolved around repetition of things like shapes, colours and sounds, but number patterns have an equally important part to play in developing children's conceptual understanding of pattern in mathematics. Rather than repeating the same numbers over and over (such as 1, 2, 1, 2, 1, 2, 1,...), number patterns are more typically sequences of numbers developing according to some rule. The simplest type of example tends to rely on generating patterns from multiples, such as 3, 6, 9, 12, 15,..., where the next term will be 18, three more than the last term. But rather than using an additive strategy, recognising the role of multiplication is more valuable in the longer term. Therefore to find the 100th term, rather than continuing the count through all the values in between, understanding that the first term is 3×1, the second term 3×2 and so on leads us to the fact that the 100th term will be $3 \times 100 = 300$. Note that there is effectively a zero term, $3 \times 0 = 0$, even though we do not see it here; starting at zero gives us the simplest type of sequence since it generates exact multiples. Note, however, that the emphasis here is not on developing multiplication skills, but rather on progression of children's algebraic thinking, gradually including patterns which, although they proceed in regular steps, do not start at zero and are therefore not exact multiples (for example, 5, 8, 11, 14,...).

Filling in missing numbers in number patterns is fairly common in primary classrooms. What would be helpful for promoting greater understanding of how sequences work is to give children more opportunities to generate their own number patterns, for example, by giving unpredictable starting points as in activity 6.7. Another approach which supports children's appreciation of what constitutes a mathematical pattern is to present them with a string of numbers and ask them to decide whether it is some sort of organised sequence as opposed to a rather random collection of numbers.

The number patterns themselves are not necessarily the starting point, however; some children's difficulties with algebra could undoubtedly be alleviated by greater use of context and imagery. Take the number pattern 5, 8, 11, 14,... You probably noticed very quickly that this pattern is going up in threes, but that it does not start with 3 and is not generating multiples of 3. Thinking back to the pattern in Figure 6.4, how easy would it be in comparison to identify the 100th term in this sequence? How the sequence works can be described in various ways, but whilst some learners might be able to describe this from just looking at the numbers, others might well benefit from some imagery or contextual information to support their thinking. Let us explore both approaches, starting by looking at just the number pattern:

Although the terms in the sequence are not the 3× table, the pattern is increasing steadily in threes and children may well spot that we seem to be one in front of a multiple of 3 (5 comes just before 6, 8 just before 9) or alternatively two after a multiple of 3 (the first term, 5, is 1 three plus 2, the second term 2 threes plus 2). If I now imagine jumping forwards in the number pattern to the 100th term, I know that this will be structured in the same way and occur one number before, or two numbers after a multiple of 3. Perhaps you already have an idea what the 100th term will be?

So how can pictures support us in being sure we have the right answer as to what that 100th term will be?

Images are a powerful way of supporting some learners' algebraic thinking, and matchstick pictures are a really simple and inexpensive way of achieving this. In Figure 6.5 we see matchstick pictures matching the beginning of the sequence 5, 8, 11, 14,.... Putting into words the changes from one picture to the next may be helpful in understanding how the pictures will progress, and maybe you already have a picture in your mind of the 100th tree? If nothing else, that it would be very tall! You have probably realised that each successive tree has an extra three matchsticks, adding an extra triangle to the height of the tree, so if tree number 1 has one triangle (3×1 matchsticks) then tree number 100 will need 100 triangles ($3 \times 100 = 300$ matchsticks). Of course we must not forget the tree's trunk; each tree uses two extra matchsticks to form the trunk, therefore our 100th term must be 302 based on the number of matchsticks required to create the 100th tree.

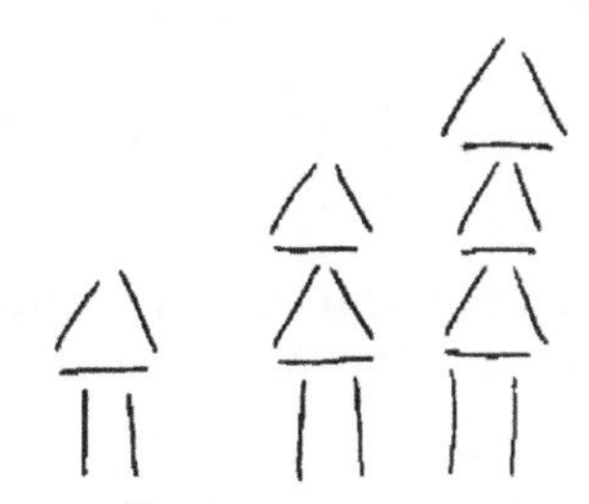

Figure 6.5 Matchstick Christmas trees

For further information on this and other matchstick patterns, see Orton (2005), but note that we do not actually have to make or draw all the patterns. The power of the imagery is in enabling the learner to understand the way in which the pattern grows and can be generalised; once we appreciate this, algebraic thinking takes over and further making or drawing may not be required.

As children grow older, they will gradually be introduced to number patterns which do not have a constant difference between the terms in the sequence. Here, too, pictures and practical activity can prove helpful.

Algebra and its ability to describe unknowns

Numbers and variables

The other main role that algebra plays is in dealing with unknowns and with values which can be variable. Whereas in the latter stages letters are used to represent these unknown or variable values, earlier examples are often of the type we see in Figure 6.6 where a box represents the number to be found.

$$\boxed{7} + 3 = 10$$

$$2 \times \boxed{7} = 14$$

$$\boxed{16} \div 2 = 8$$

Figure 6.6 Early algebra in relation to calculation

Common misconception

$$'3 + \boxed{8} = 5'$$

The person completing the above calculation has treated the box for the unknown more like an invitation to write the answer in it, overlooking the need for equivalence.

All the examples in Figure 6.6 have something in common: definite values satisfy the boxes. This is in direct contrast to Figure 6.7, where a variety of answers can be suggested to satisfy each calculation. Even drawing on just positive whole numbers, an infinite number of pairs are available with a difference of 5, suitable therefore for completing the subtraction example. If we expand our thinking to include negative numbers and fractions or decimals, then the same becomes true of 'something plus something equals 10'. The term 'variable' is used to describe values that can change.

$\square + \square = 10$

$\square - \square = 5$

$\square \times \square = 36$

Figure 6.7 Calculations involving variables

Thinking in terms of variables and unknowns can be introduced to children at a young age through consciously offering activities and asking questions with an element of uncertainty. Making up a story about going shopping one day and buying lots of tins of soup, then going shopping again the next day and only buying a few incorporates no actual values, but could initiate discussion of possible numbers of tins. If nothing else, we know that the number of tins on day two should be fewer than on day one! You may want to look back at activity 3.17; this was an activity requiring algebraic thinking in the context of difference.

Other opportunities for generalised arithmetic present themselves where children are building up collections of facts (see, for example, activity 7.18 in the next chapter). Using the approximate relationship between miles and kilometres, I have begun to complete the table we see in Figure 6.8. When choosing where to go for a walk I prefer to know what it will be in miles so if it is presented in kilometres I may want to convert it. Whilst some of the relationships are known facts (the obvious ones), the final column gives me a way of converting the distance, however many kilometres it is. Multiplying by 0·625 is equivalent to finding five-eighths of however many kilometres, since a mile is longer.

km	4km	8km	16km	?	n
miles	?	5miles	10 miles	20 miles	0·625n

Figure 6.8 Building the relationship between kilometres and miles

Notation of algebra

Continuing the theme of unknowns, we will now explore a simple story, thinking about how algebraic notation could be used to describe it. Imagine a bag of marbles; this bag, like all the others in the story, contains 10 marbles. If I go to the toy shop and buy three bags, I would have 30 marbles in all, and this can be recorded using regular arithmetic notation: $10 \times 3 = 30$. But what if I go to the shop with my life savings and buy all the bags of marbles in the shop – how many marbles will I end up with then? This is more challenging to record since there is now a variable involved; the shop could have lots and lots of bags of marbles or only a few bags, we just do not know what we will find when we get there. The solution to our recording conundrum lies with algebra, using a letter to represent the unknown number of bags; any letter will do but I will use b as it will remind me that I am thinking about not knowing how many bags of marbles the shop will be able to sell me. Once I find out, a calculation of $10 \times b$ (or more conventionally $10b$, losing the multiplication sign) will tell me how many marbles I have bought: 10 per bag, multiplied by the number of bags.

It should be noted that whilst I have used a related letter in my story (b to remind me I was thinking about numbers of bags), the letter does not represent a 'thing' (the bag) but a variable, the as yet unknown *number of* bags. A value which could be large, small, or even zero if the shop is out of marbles! If children come to see the letter as an abbreviation and interpret $10b$ as shorthand for 'ten bags', this can have unfortunate repercussions for later algebra; encourage them therefore to articulate it more accurately. For this reason some teachers stick to tradition (x and y) or use n for unknowns; however, if children only come to associate certain letters with algebra, then the eventual use of others may result in confusion!

Occasionally there are some loose marbles in my story; imagine I happen to have two in my pocket. How would this affect the number of marbles I end up being the proud owner of? We still do not know how many bags of marbles the shop has in stock, but we now know that whatever number I can purchase, I will have an additional two marbles. These are referred to as the 'constant', in this case a constant of + 2. So my total number of marbles on this occasion is going to be $10b + 2$. Note the multiplication element takes precedence over the addition (or subtraction) of the constant. In this context subtraction would be needed if we imagined a scenario in which I owe a friend three marbles and cannot give them their marbles until I have been to the shop. In this instance the total number of marbles will be $10b - 3$ at the end of the story.

Just as we can notate the marbles story, we could also have notated the generalisation of the Christmas tree pattern. If we were asked about the number of matchsticks for *any* tree (this is the variable aspect), we could plug that number into $3n + 2$ and it would tell us how many matchsticks were required. Hence the overlap between the two aspects of algebra: pattern and unknowns.

Incidentally, the principle of variables and constants lies behind the calculation of many utility bills. For example, my telephone company charges me a set amount per month for line rental, and on top of this I pay so much per call depending on the type of call and how long it lasts. This is the variable bit since the calls that will be made are not known in advance; the line rental in contrast is a fixed charge, the constant element of the algebraic calculation.

The examples so far have involved variables in relation to calculation, but they are also found in measures, typically as an aspect of geometry, for example, to label unknown lengths or angles. See, for example, activity 6.15 which provides a nice opportunity for children to use their reasoning skills. $3a + 4a + 5a = 12a$, but remember that a in this case represents an unknown or variable length which could be any number of any unit. Without an actual value a calculation such as this one cannot be taken any further, it just makes a statement about a relationship between the lengths. For some children this takes a bit of getting used to!

Common misconception

'$3a + 4a + 5a = 12a$ can't be the answer, it's not finished'

Discussion can help children to appreciate that rather than being unfinished business, algebraic statements are a way of writing mathematics that will work for lots of different numbers, big and small!

Algebra through problem solving and investigation

In this final section we will explore the potential of problems and investigations to develop children's ability to generalise and cope with variables, starting with some families of animals made from Multilink. I am indebted to Andrews and Sayers (2003) for convincing me that more was possible with younger children. Having introduced the children to the first Multilink dog in the family, the size of dog was gradually increased (see Figure 6.9), and children were given plenty of time to engage with the dogs through practical activity. Their attention was later drawn to the properties of the Multilink dogs, emphasising those features which stayed the same (which we might call 'invariants') and those which increased each time (the 'variant' features). Ability to distinguish between the two is key to being able to identify patterns and to turn them into generalised algebraic statements.

Figure 6.9 Family of Multilink dogs

We have met the idea of variant and invariant features before of course; whilst the matchstick trees got taller and taller, the trunk stayed the same, requiring just a constant two matchsticks each time. In the case of Multilink animals it can be heads, shoulders, tails,... staying the same whilst legs, necks, arms, bodies,... are getting steadily longer. To support children in spotting these features we might advocate some use of colour; perhaps not to start with, but at some point later when there has been an opportunity for discussion of what is changing and what is staying the same. My rationale for suggesting use of colour to distinguish variant properties from invariant properties later rather than sooner is that people do not always spot the growth of a pattern in the same way. This is discussed further in activity 6.9 where there are plenty of opportunities for algebraic thinking! The sequences of numbers have been tabulated as this can be helpful to some learners in making the patterns more obvious. Graphing can also be a useful way of illustrating the relationship between the model number (typically shown on the horizontal axis) and the number of cubes. The line of the graph reminds us that, by extending the line, we could have larger or smaller values (including negatives) and that other values exist on the line, such as decimals. Having said that, with something like Multilink animals it is hard to think in terms of negative or half dogs!

In Figure 6.10 we see two examples of another style of problem which lends itself to algebraic thinking. The second one is from Potter (2006: 174). What both problems have in common is that we are told the overall totals, but do not know how the total is split in each case; in other words, we are working again with unknowns. Given only part of the information from either problem these would remain unknowns, but in each case the second sentence provides us with enough additional information to find a solution.

I buy a mixture of teas and coffees, five in all, and it costs me £10·50.	The teas (£1·90) were cheaper than the coffees (£2·40) so how many of each?

There were a mixture of frogs and princes in a palace; there were 35 in total.	Between them they had a total of 94 feet so how many of each were there?

Figure 6.10 Example problems

This style of problem becomes more algebraic at a later stage through solutions being sought via simultaneous equations. I will leave those to my secondary colleagues, but primary teachers can certainly provide children with valuable experience by incorporating problems of this type within their teaching. So how do we solve them? How do we find out the number of teas versus coffees and frogs versus princes? The various children with whom I have tried the frog and prince problem often draw on their awareness of pattern to work towards a solution. They often start with a guess and seek to improve it based on having too many heads, too few feet or whatever, thus the patterns of the pairs of numbers go up and down. Or sometimes children start at one extreme and build the pattern from there:

- If all 35 were princes, there would only be 70 feet.
- If there were 34 princes, there would be one frog and that makes 72 feet.
- If there were 33 princes…

Other children have been known to draw pictures and to solve the problem very successfully in that way. Whatever the method, workings tend to be personal jottings, though sometimes children tabulate the numbers in some way. Clearly they do need to know the numbers of legs a frog has to be successful and this is not always the case!

Algebra activities

All the activities described in this chapter are designed to get primary-age children thinking algebraically, in particular, pattern spotting and generalising. Encouraging them to articulate what they notice is a crucial part of building secure understanding. Regular worksheets, such as those which ask for missing numbers or for patterns to be continued, can of course be used, but think about what they offer in terms of opportunities to generalise. Remember activities can almost always be adapted, particularly where you feel different numbers or contexts might be more suitable for your class. There are usually simpler or more challenging patterns or sequences that can be substituted for the ones I suggest.

Pupil activity 6.1

The children as patterns
Y1 to Y2

Learning objective: To begin to appreciate repeating patterns.

Something as simple as the children lining up can be used to generate repeating patterns (boy, girl, boy, girl, or colours of socks or whatever). Take those opportunities that come along in odd moments to create patterns; look at them, listen to them (saying them out loud), think about who would come next if we added another child to the line. Bringing pattern to the children's attention on a regular basis will help to prepare them for pattern work in dedicated lessons.

Pupil activity 6.2

Practical patterns
Y1 to Y3

Learning objective: Ability to articulate repetition in pattern.

In activity 6.1 the emphasis was on the teacher generating the patterns, whereas here the onus is on the children who can use whatever materials are to hand in the classroom. Introduce the activity by putting into line a few very random items and asking the children to decide whether there is a pattern; hopefully the need for elements to be repeated will be part of the discussion. Put this to the test by asking pairs to make repeating patterns; this might be with toy vehicles in a traffic jam, items from the home corner, counters – it really does not matter. Photographing the patterns provides both a good record and an opportunity to then talk about the patterns in terms of what would come next and predicting further ahead.

Pupil activity 6.3

Rules for patterns with two variables
Y2 to Y4

Learning objective: Ability to articulate the 'rules' for a pattern involving two variables.

As children become more competent at working with patterns and recognising units of repeat, they can work with more than one variable; in this activity the children would work with both shape and colour. If children make their patterns around the edge of the page, the repeats can then join to form a continuous loop and there is space in the centre for the child to describe the rule for the pattern. The explanation can tell us a great deal about their understanding of generality.

Here we see an example; a child whose response is reasonably economical and concise, in contrast with others along the lines of 'My pattern goes red square, yellow triangle, red square, yellow triangle, red square,…'.

My pattern are square
triangle and my triqngle
is yellow and my
sqqare is red.

Pupil activity 6.4

Musical patterns
Y1 to Y4

Learning objective: Ability to identify a unit of repeat in music.

Simple rhythms offer a really nice way of working with pattern, and might start with sequences of claps or notes with the children trying to isolate the unit of sound that is being repeated. The children can then be given the freedom to generate the simple pieces of music by playing the group's chosen percussion instruments repeatedly in a particular sequence, making a record of the pattern. In the long term this might mean exploration of conventional musical notation through which things like tempo can be taken into account. But to start with it is about agreeing symbols that will represent the sounds so that they can be written down.

This series of pictures told the group that the triangle would be 'tinged' three times before a couple of hits of the claves and finally a clash of the cymbals!

Pupil activity 6.5

Units of repeat
Y2 to Y6

Learning objective: To develop awareness of the size of the repeated unit in a pattern and the relationship between this and prediction.

This activity assumes the children are reasonably familiar with the sorts of patterns discussed so far and aims to draw the children's attention to the size of the unit of repeat; in other words, to how many terms there are before a pattern starts to repeat again. Discuss a simple example such as the one we see here which repeats after three pictures.

Prepare lots of different examples with different size units of repeat for the children to sort collaboratively into labelled sets. Encourage discussion to resolve any disputes about where a pattern belongs and aim to include some challenging examples, for instance, by using something like an arrow in different orientations.

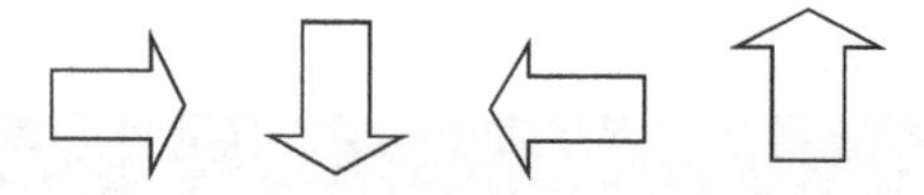

Whilst such patterns can then be used to think about what comes next, begin to challenge the children to predict further and further ahead. Encourage the older children to verbalise how they know which symbol will occur at different points in the pattern. The activity therefore provides a good mechanism for assessing whether a child has some understanding of the role of units of repeat in generalising.

Pupil activity 6.6

Counting patterns
Y1 to Y6

Learning objective: Ability to describe what happens to a number sequence and to identify missing numbers.

As children grow older they will be used to the counting numbers and will develop familiarity with other number patterns (such as 2, 4, 6, 8,...and 10, 20, 30, 40,...), noting that these are sequences rather than repeating patterns. Be ready to discuss this with the children if you think making this distinction more explicit would help. Familiar sequences (offering links with multiplication) can be presented with missing numbers for the children to identify, but to develop the children's algebraic skills the focus then needs to be extended to discussion of how the pattern grows, beginning to think about the sorts of numbers that will appear further ahead. Rather than familiar number patterns, older and more able children can be asked to discuss patterns such as

147, 247, 347, 447, 547,...

with the potential of generalising in words or by creating an algebraic statement. The 10th number will be __. I know this because __. I could find any number in the sequence by __.

In addition to patterns which proceed forwards in equal steps, incorporate some which go backwards, some which grow in a non-linear way (for example, doubling) and some which involve fractions or decimals.

Pupil activity 6.7

Understanding pattern generation
Y2 to Y6

Learning objective: To develop a creative approach to pattern generation and the ability to describe the pattern.

In activity 6.6 the teacher provided the number patterns for the children to fill in missing numbers and discuss the patterns; here they will be generating their own from unpredictable starting points, such as 6, 3,... or 2, 4,... This is intentional! Given the opportunity to explore where a pattern *might* go supports the ability to identify where given patterns actually go. Sharing ideas is of course the vital ingredient, otherwise the children will not necessarily realise that others have let the starting point lead them off on a different pattern. How many different examples can the children come up with for the same starting point? If no one mentions negative numbers, fractions or decimals, you might want to propose these as possibilities if you feel they are suitable for the age group.

Another way of generating the patterns is for children to pick two number cards (or roll two dice) and to decide how to incorporate those numbers into a pattern.

The same unpredictable style starting points work for other patterns, too, of course; how might circle, square,... carry on?

Pupil activity 6.8

Random or pattern?
Y2 to Y6

Learning objective: Ability to distinguish between a recognisable sequence of numbers and a random collection with no discernible pattern.

Presenting children with strings of numbers and asking them to decide (ideally collaboratively, to encourage justification of decisions) whether or not they are organised sequences of numbers is another good way of enhancing children's appreciation of pattern in mathematics. Once the children have sorted the strings of numbers, make sure you listen to their reasons for those they believe to be sequences even if you're not sure you agree. They catch me out sometimes when they spot a potential pattern that I had not thought of!

Pupil activity 6.9

Multilink growth
Y2 to Y6

Learning objective: Ability to interrogate properties to identify similarities and differences between sequential Multilink models.

Earlier in the chapter we met the idea of Multilink animals; here we explore patterns generated in the context of window frames, but it could as easily be furniture (beds, chairs, tables, increasing in size) or any type of Multilink shape, not necessarily contextual.

All the window frames are square and the first requires 8 cubes, the second 12, the third 16, and so on. As the frames increase in size so does the hole in the middle you can see through, though theoretically there's a 'zero' frame with just four corners and no hole!

Number of window frame:	0	1	2	3	4
Number of cubes needed:	4	8	12	16	?

Once the children have had a chance to use the Multilink to produce some of the early frames and note the number pattern, it is time to start attending to the properties of the window frames. At this point colour might be used to pick out the features. The long-term aim of course is that the similarities and differences between one model in the sequence and the next enable the children to generalise to such an extent that they can predict further ahead. How many cubes will be in the 100th frame? How do you know? What I have attempted to show with use of shading is that people might interpret the properties differently in terms of grouped or individual cubes. How could the features of each of these be applied to the next window frame in the sequence and how useful would it be in helping you to find a general statement?

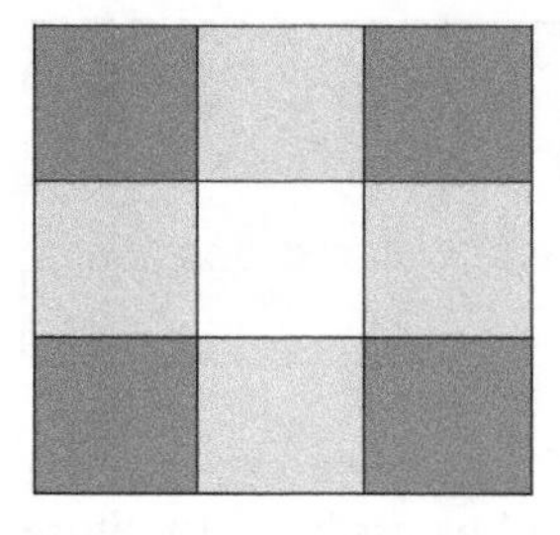

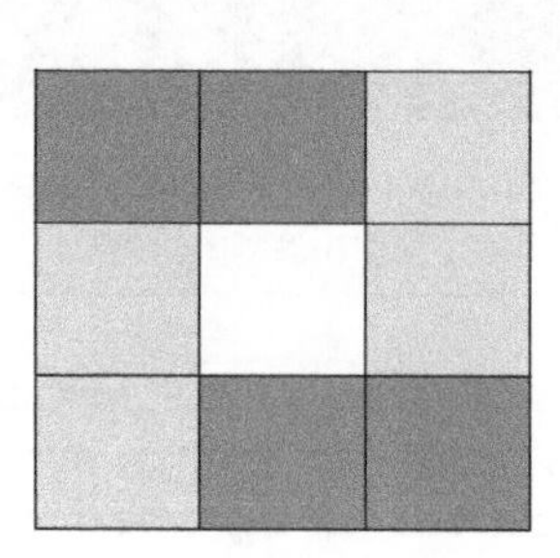

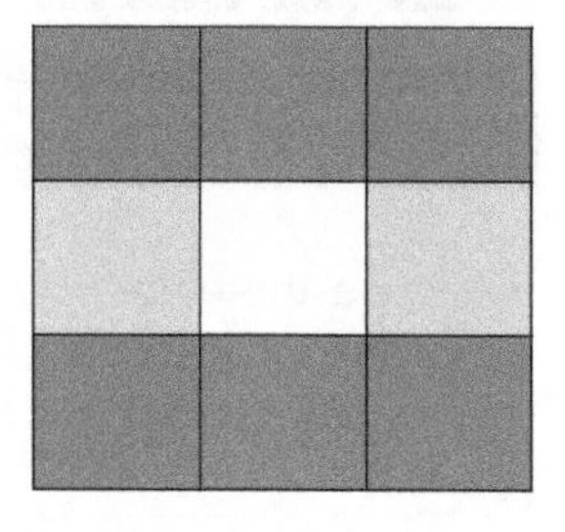

The first picture works well for me in terms of generalising. It has one light grey square along each side and a dark grey square at each corner; as this is window frame number one, the 100th frame will therefore have 100 light grey squares along each side (400) plus the four corners: 404 cubes in all. We might write this as $4n + 4$ which is equivalent to $4(n + 1)$, which could represent the second image where we see four lots of two. Why $n + 1$? Well, we are looking at a picture of frame number one but this pictorial interpretation is built of twos. The third possibility is a mixture of ones and threes and could be written algebraically as $2(n + 2) + 2n$ which is equivalent to $4n + 4$. Try popping 100 into each statement as the variable if you want to check; the answer is the same every time! Even by the end of the primary age phase we might not get to the stage of algebraic statements with all children, but opportunities for them to spot and talk about the patterns in their own way are a good start.

There is at least one picture missing, by the way. Thinking about numbers of cubes needed for window frames *without* versus *with* holes in the middle might suggest yet another way of generalising how many cubes are needed. Test yourself at the end of the chapter!

Pupil activity 6.10

Jubilee street party
Y4 to Y6

Learning objective: Ability to turn a given sequence into a general statement that will work for any value.

Various authors (such as Haylock, 2010) document table arrangements as a good way of generating patterns to be generalised, and I can attest to this myself, having used it as a starting point for years! Whilst we could use it with younger children, it is perhaps best used with the older ones with the intention of working towards an algebraic statement.

Imagine that we are to have a party to celebrate the Queen's Diamond Jubilee and, in true street party style, we want to lay lots of tables end to end. Each table can normally seat six: two on each side and one at each end. Modelling this with my classes, I usually pretend that one of the folk at the end will have to keep getting up and moving as more tables are added. Drawing attention to the end folk is beneficial since these two are essentially the constant; however long a line of tables we manage to achieve, there they will be, sitting at either end. (Incidentally, if we had no table they'd still be our constant; sitting there knee to knee!) Drawing pictures like the ones we see here is the typical way to investigate this problem further, but making it clear to the children that what we are after is a way of finding out how many people can sit down whatever the number of tables. I typically throw out some random numbers (what if there were 53 tables?) to judge which children are beginning to understanding the problem in general terms.

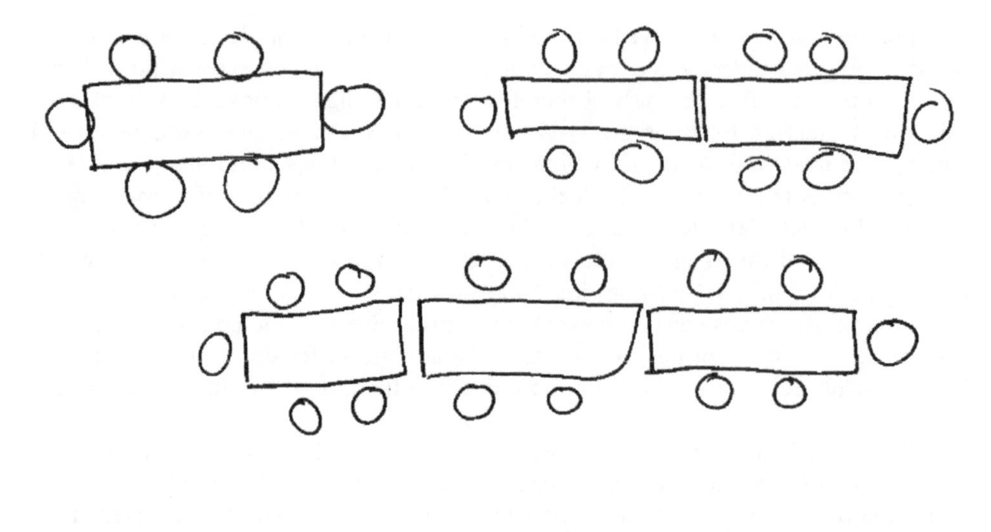

Number of Tables	1	2	3	4	5	6	7
Number of People	6	10	14	18	22	26	30

Having drawn several pictures this child recorded further results in a table of values; he went on to say that 402 people could sit down if there were 100 tables, articulating clearly how he knew. This could then have been translated into an algebraic statement in general terms.

Pupil activity 6.11

Matchstick algebra
Y5 to Y6

Learning objective: To appreciate the relationship between a number pattern and a physical or visual pattern.

Matchstick patterns provide an alternative to the table arrangement idea in activity 6.10. Having shown the children some basic sequences like the Christmas trees we saw earlier in the chapter, give them some time to explore the sorts of picture sequences they could make with the matchsticks. How many matches do they need in each successive picture? How far ahead can they predict?

Following the open-ended start, I would then show the children some number sequences, giving them time to talk about how they see the number pattern developing and whether they can predict some of the numbers which will occur in the pattern and explain how they know. Can they then suggest a picture sequence to accompany it? Approaching it this way round and giving the children the chance to suggest the matchstick arrangements really tests their understanding of the relationship between the two. For example, the sequence 3, 5, 7,... might result in triangles getting gradually taller and pointier as we see here.

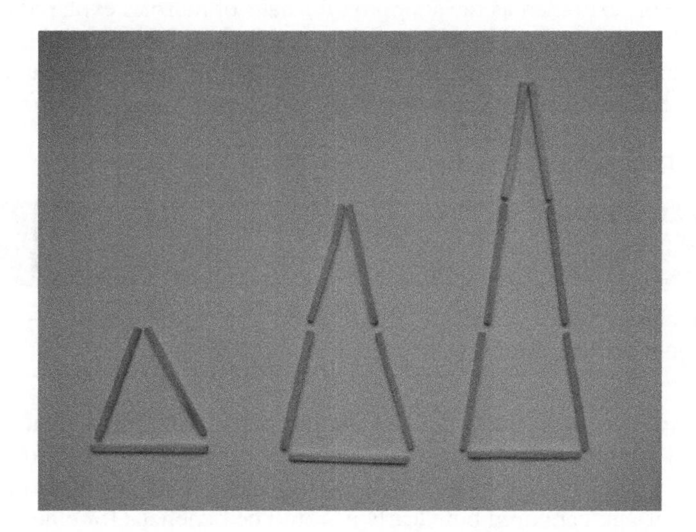

The level of support throughout this lesson would need to vary, depending on the children's responses.

Pupil activity 6.12

Variables
Y1 to Y6

Learning objective: To develop appreciation of variables in calculations and ways of recording situations involving variables.

Even with the youngest children work can be done to encourage thinking in terms of variables – those occasions when we cannot be sure about the actual numbers involved. Telling shopping stories where you cannot quite remember what you bought provides the starting point for this activity which relies on secure number bond knowledge. For example:

- I bought five tins of soup; some were tomato and some were chicken, but I can't remember how many of each. How many of each could it have been?

Scenarios such as this form the basis for thinking about the possibility of different answers and resources could be used for illustration purposes, opening the bags to see who was right once the children have had an opportunity to make their suggestions. This activity can be related to recorded mathematics looking for different ways of finishing 5 = ? + ? and other calculations. Eventually, moving beyond whole numbers, answers to questions like this one invite an infinite number of responses!

As the children's understanding of variables gets stronger, they begin to explore unknowns in contexts such as the shopping for bags of marbles explored earlier in the chapter. Rather than recording the possible calculations, encourage the children to record general statements. Remember that as well as the teacher making up algebraic stories, children could be given opportunities to make up some akin to the marbles one.

Pupil activity 6.13

Think of a number
Y1 to Y6

Learning objective: To appreciate the effect of the same operation(s) on different numbers and know how to reverse the operation(s).

You may have come across the idea of function machines in mathematics. We put a number in; something happens in the 'machine' and out comes a different number (or perhaps the same one if whatever happens leaves it unchanged). I have used the idea with infant-aged children but clearly it would be beneficial for older children too, since the numbers and the operations can become more challenging. The level of challenge is in part down to whether it is a single operation (just one thing happens) or a combination of two or more (perhaps separating out the operations rather than writing them jointly as we see here). The operations and starting number can be known or unknown, so the child might be trying to work out what the machine is doing or what number was put in. After this children can explore the possibility of finding a machine which can reverse a particular sequence of operations, as illustrated in the pair we see here.

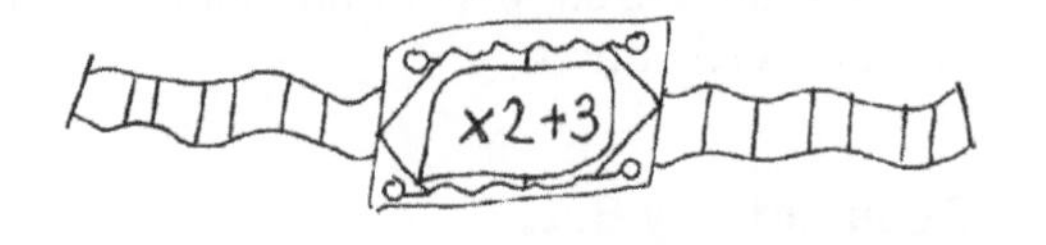

Working without the function machine context is also possible through what are sometimes called 'think of a number' activities. Participants can think of any number (a variable, n) and a particular sequence of operations happens to it:

Think of a number. Double it. Add 2. Halve it.

Recording this using algebraic notation we can think of this in terms of $(2n + 2) \div 2$ and you might notice something interesting happens with each value you try (and that the algebra can be simplified in keeping with this). As with the function machines, we can also think about how to undo the operations. How might the children describe this using algebraic notation? Activities such as these provide support for children with regard to ways of solving equations.

Pupil activity 6.14

Using both bits of information
Y3 to Y6

Learning objective: To develop strategies for tackling problems where the information is given in two parts.

Problems like those discussed earlier in the chapter (about the frogs/princes and teas/coffees) help to lay solid foundations for later work in algebra and also help to develop children's problem-solving skills. Remember of course that one individual's approach will likely vary from someone else's but that children will benefit from sharing ideas. Two more examples are given below, Rather than giving both parts at once, giving one part at a time can help children to picture the situation and begin to consider some possibilities.

In a farmyard there are some chickens clucking around and some pigs; there are 29 animals in total.	The total number of legs was 78. How many chickens and how many pigs were there?

8 people were sitting in a room. Some were sitting on chairs with 4 legs and some on stools with only 3 legs.	Counting all the legs (people, chairs and stools!) they had 45 legs between them. So how many people were sitting on each type of seat?

Pupil activity 6.15

Reasoning about unknowns
Y5 to Y6

Learning objective: To understand the role letters play in representing unknowns in geometry.

Some years ago now I was disheartened to experience some algebra homework which rather than encouraging James to use his powers of reasoning, seemed to expect him to suspend them! (See Rickard, 2010.) This activity is hopefully based on doing a better job of encouraging reasoning. Give children pictures of the type we see here and ask them to look and think carefully as you suspect there's something not quite right.

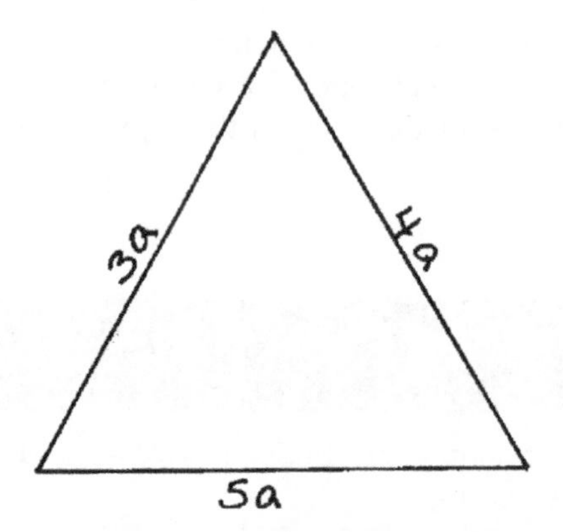

Perimeter = $3a + 4a + 5a = 12a$

Hopefully the children will query the fact that the unknown lengths are at odds with the picture, which appears to be an equilateral triangle. However big or small a value we pick for a, the lengths should be all the same, whereas given the labelling of this picture we might reasonably expect all the lengths to be different. Ask the children to suggest a more suitable triangle for these lengths.

A nice extension challenge is to ask, if the sides of the triangle had instead been $3a + 4b + 5c$, whether there are there any values for the variables which *would* make the triangle equilateral.

Letters can of course also label unknown angles and my desire for good-quality reasoning opportunities remains the same; expect children to make sense of their mathematical experiences!

Pupil activity 6.16

Next door numbers
Y5 to Y6

Learning objective: To understand that $n - 1$ represents one less than a number n, and $n + 1$ the value one more.

Adding three consecutive numbers gives rise to a rather nice investigation which is algebraic in nature. Differentiation can be easily achieved by giving a range of triples for the children to pick from; once they have investigated a couple, they can then try others of their own choosing.

9, 10, 11

50, 51, 52

4, 5, 6

1, 2, 3

99, 100, 101

6, 7, 8

What you may have noticed if you have tried this yourself is that the totals are always three times the middle number. Algebra helps us to explain why this is the case; if we call the middle number n, the value before it is $n - 1$ and the value after it $n + 1$, so adding these values, $(n - 1) + n + (n + 1)$, gives us $3n$ as the -1 and $+1$ cancel each other out. Clearly, rather than telling the children this, you will want to give them plenty of time to work at this investigation, talk with each other and reflect on why the totals are as they are.

Summary of key learning points:

- algebra is related to pattern and these can be either repeating patterns or form part of a sequence;
- in-depth appreciation of pattern underpins much in algebra, and sense of pattern enables us to make general statements about elements at any stage of the pattern;
- images, sounds and practical activities support some learners in appreciating pattern;
- ability to articulate what we notice about patterns develops towards conventional notation of algebraic statements;
- algebra allows us to deal with unknowns and variables, and this has real-life application.

Self-test

Question 1

What is the 100th term in the sequence 2, 3·5, 5, 6·5, 8,...? (a) 152, (b) 1505, (c) 150·5

Question 2

How could we write the *n*th term (in other words, any term in the sequence that anyone cares to pick) of the sequence in question 1 algebraically? (a) $1{\cdot}5n + 0{\cdot}5$, (b) $0{\cdot}5n + 1{\cdot}5$, (c) $2 + 1{\cdot}5n$

Question 3

For the Multilink window frames in activity 6.9 another generalised statement is (a) $(n + 2)(n + 2) - n^2$ (b) $(n + 2)^2 - n^2$ (c) $(n + 2)(n + 2) - (n \times n)$

Question 4

The recommended ratio for the numbers of adults (a) to number of Guides (g) for unit meetings is one adult to every 12 girls, so $a : g = 1 : 12$. Which of the following expressions expresses the relationship between Guides and adults? (a) $12g = a$, (b) $g = 12a$, (c) $g = a \div 12$

Self-test answers

Q1: (c) Once we have identified that the terms are going up by 1·5 each time it helps to recognise that the first term, 2, is $1{\cdot}5 + 0{\cdot}5$. The 100th term will therefore be $(1{\cdot}5 \times 100) + 0{\cdot}5 = 150 + 0{\cdot}5 = 150{\cdot}5$.

Q2: (a) Given the explanation for question 1, the *n*th term will be equal to $1{\cdot}5n + 0{\cdot}5$.

Q3: (a, b, c) All three answers are actually equivalent expressions. If we think about a solid square of Multilink (window frame, no hole) and use our knowledge of multiplication and area, this gives us the $(n + 2)(n + 2)$ or $(n + 2)^2$, but we then have to subtract a smaller square (n^2) to get our hole back!

Q4: (b) The answer might well surprise you and you would be in good company if you chose answer (a)! I used the letters g and *a* on purpose – but remember, these are not abbreviations for 'girls' and 'adults' but are unknown numbers: we can only guess how many girls or how many adults will come to the meeting. A good way of checking for the right algebraic form is to try some numbers. If we have four adults the original ratio information suggests we can have 48 Guides; sure enough, if we put the numbers into equation (b), this is the answer we get. Note that fewer than 12 Guides would give us an answer of less than a whole adult; algebra does not always help us in reality!

Misconceptions

'3 + [8] = 5'

'$3a + 4a + 5a = 12a$ can't be the answer, it's not finished'

7
Measures

About this chapter

This chapter describes the common areas of measurement that primary children will engage with. Much of what teachers and children need to understand is generic across areas of measurement and is dealt with under the following headings:

- The role of units in measurement
  - From non-standard to standard units
  - The relationship between measurement and number scales
  - Compound measures
  - Ability to estimate measures
- The role of language in measurement
  - The language of comparison
  - The language of length
- Specific relationships and differences in measures
  - Area and volume
  - Volume versus capacity
  - Weight versus mass
  - Time

What the teacher and children need to know and understand

The role of units in measurement

One of the most important things to recognise about the teaching of measurement is that there are just so many different ways in which all manner of things can be measured and assigned a value of some sort. 'How heavy is it?', 'How tall are you?', 'How fast is it going?' and 'How hot is it?' are all examples of questions related to measurement. As we will find out in this chapter, many of these aspects are linked and thus learning about one area of measurement may well help children to understand another; this is particularly the case when learning about how units operate. Much of our teaching will involve practical activity, through which children will eventually learn about metric units, but are likely to start with non-standard units. Of course we do still have various other measures, such as pounds and ounces, in common enough use that we would be doing children a disservice not to recognise these in our teaching too.

From non-standard to standard units

Progressing towards the use of standard units, whether it be metric units such as litres, imperial units such as pints, or units of time, children have to be helped to appreciate the need for standardisation. Most teachers explore this with young children through practical and discussion activities where the use of non-standard units results in the same item apparently being a different size. 'I measured the length of the carpet in footsteps and I think it's 13 footsteps long. You say it's 18 footsteps long. How can that be right?' Quality discussion may help the children to appreciate the need for units such as metres if they are getting different answers for the length of the same carpet area measured in footsteps.

Whilst footsteps for length, or handfuls of conkers balancing another item, are totally non-standard, there is an interim stage we might consider: the use of equipment which is itself standard, something like Multilink being one of the likely resources to hand. Measuring how long something is, or how heavy, we ought now to be getting the same answer, helping to exemplify the way in which standard answers are achieved. Through planned teaching we also need to focus the children's attention on measuring skills – for example, how to line the Multilink up with the start of the object I want to find out the length of, making sure I do not leave different size gaps between the cubes, and counting them carefully.

In your teaching you will also want to address the possibility of partial units when the need arises; once children start getting involved in measuring activities, you will quickly find that this occurs and we start to get answers like 'my shoe is a bit longer than 9 Multilink' or 'this teapot holds almost 5 cupfuls'. Before children are familiar with decimals or fractions they can be encouraged to share their own ways of verbalising their findings: '3 and a bit', 'not quite 4', 'between 3 and 4'. You may well find children using ‘and a half’, which will provide an ideal opportunity to discuss what is meant by this and, given time, will lead towards the relationship between the language, the symbol $\frac{1}{2}$ and the decimal equivalent once we are using standard units (3·5 cm). The early experiences help to build towards the notion that measurements are only ever approximate; we are dealing with something that is continuous and the value we assign is only as accurate as the tool in use allows. Take your average kitchen jug and try to measure out, say, 34 ml – not an easy task!

The relationship between measurement and number scales

Once children appreciate the need for standard units and have developed some of the early skills associated with measuring, they can gradually apply the knowledge and skills to use measuring equipment such as rulers, jugs, weighing scales and protractors. Ability to interpret a number scale is a crucial skill here, so some reinforcement of the number system is recommended alongside measurement work, including both ordering and positioning of numbers and multiples.

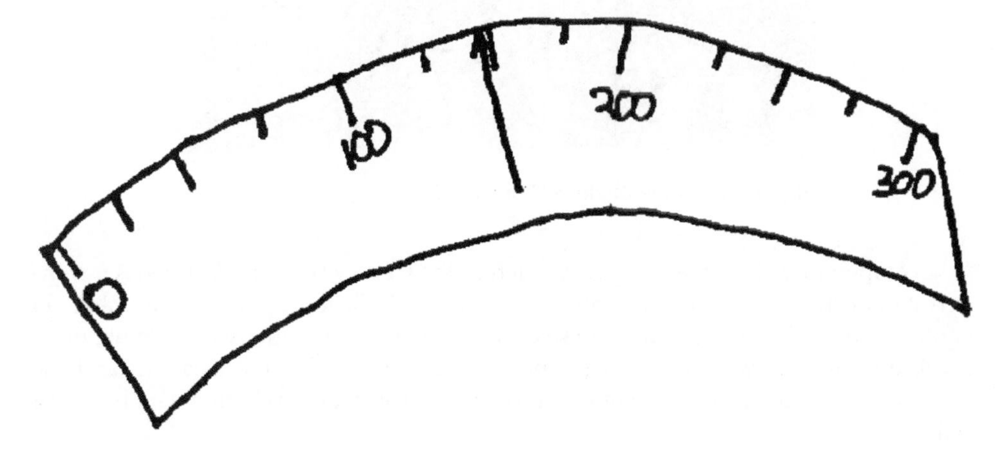

Figure 7.1 A child’s number scale

A secure grasp of the number system will allow children to appreciate all the numbers that would appear between 100 g and 200 g in Figure 7.1, and potential values in various positions, such as that marked by the arrow. We might consider the unlabelled divisions (which we note are equally spaced), maybe the half-way marks, the way in which the remaining gaps have been subdivided, and so on. In this case the scale is going up in multiples of 25, so ability to count in 25s will be an advantage here, whereas on other occasions it might be fives (take the minutes on a clock face), or 50s or 20s. Strong number skills therefore support much of our work in measures. Conversion between units similarly requires sound number skills, in particular those associated with place value, and multiplication and division by multiples of 10; many a time noughts are lost or gained when swapping between millimetres, centimetres, metres, kilometres and so on.

Angles can be measured using a protractor (in degrees with the symbol °), but an added complexity here is that the number scale on many protractors runs in both directions, regularly resulting in predictably incorrect answers. Key facts such as 360° being a full turn are useful, and compass points are related to degrees. For more on aspects of angle, see Chapter 11.

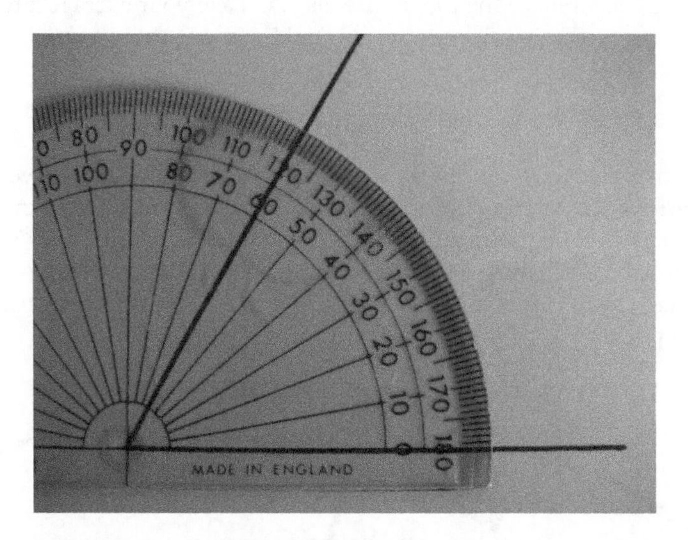

Figure 7.2 Is the angle 60° or 120°?

Temperature also utilises degrees (such as the Celsius scale) and provides a wonderful opportunity to focus on a number scale which illustrates negative numbers in context and can be linked to work in science and geography. Links with data handling are also ideal since we can access temperature charts for different locations and use these to explore the measurement of temperature as well as developing data handling skills.

Common misconception

'12°C is twice as hot as 6°C'

Whilst on the topic of measurement scales, it should be noted that some work slightly differently to others. Most of the ones we are familiar with are ratio scales where the relationship between the values means we can safely say that a 12 cm piece of ribbon is twice as long as a 6 cm piece, and if we jokingly say we have a 0 cm piece we don't have any ribbon! This does not work, however, with something like temperature which is measured on an interval scale. We are unable to claim that 12°C is twice as hot as 6°C, and 0°C does not suggest an absence of temperature!

Time should be considered more carefully than other aspects of measures since it draws on two different scales. Like temperature, we know that time of day is an interval scale because whilst I can say that 2 o'clock is four hours earlier than 6 o'clock, 6:00p.m. is not three times 2:00p.m. However, if I am talking about lengths of time, time works on a ratio scale; something can take me twice as long, or only a third of the time.

One final type of scale is an ordinal scale which is only useful for ranking, for example, the number of stars awarded to a particular hotel or a child being asked to rate their enjoyment of something using smiley faces (see Figure 7.3). Here, too, we cannot claim relationships between the values: a four-star hotel is not twice as good as a two-star establishment! Nor should we calculate averages from such data. For more detail on the topic of measurement scales, see Haylock and Cockburn (2008).

Figure 7.3 An example of an ordinal scale

Compound measures

Late in the primary phase children will begin to appreciate that occasionally measuring requires a combination of two different units, and this pairing of units indicates compound measures. Speed is an example of a compound measure since here we need distance and time, as in miles per hour, to denote how fast something is travelling. Another is density, where I am thinking about how heavy a certain volume of something is: lead is very dense at more than 11 g/cm^3, whereas aluminium is less dense at only about 2.7 g/cm^3. Pressure also involves mass, but this time in relation to area, with units like g/cm^2. Compound measures are particularly useful in relation to understanding aspects of the science curriculum.

Ability to estimate measures

Once working with standard measures, children need lots of hands-on experience if they are to develop an ability to make sensible estimates. Some aspects of measurement, such as length, are easier than others to estimate. Mass is one of the more challenging, although abilities do of course vary from pupil to pupil. Having been offered lots of hands-on opportunities, children are more likely to get used to the feel of how heavy a certain number of grams are, how long they think something took, how much they think something holds and so on. Seeking consciously to improve estimates is important; children's first guesses may be wildly out, but they should be encouraged to think about how one answer might influence their next estimate.

Effective teachers will also want to consider the everyday application of estimation skills so that teaching can better reflect reality. Central to this is being able to distinguish between those occasions when accuracy is important and we might not want to rely on an estimate (taking medicines, for example), and those times when a reasonable guess will be good enough. Leading on from this are the practical implications of whether we might choose to over- or underestimate in different situations. Cookery offers us many opportunities to consider these issues, such as not adding too much curry powder; I can always add more but will struggle to take some out! All these reflect the reality of what measurement is all about and if we teach children the skills without getting under the skin of how the skills are actually used in real life, then we are only doing part of our job.

Common misconception

'This has long "arms" so it must be a big angle'

When estimating angles we know that some children are influenced by the length of the 'arms' and will need to come to understand that this feature does not affect the size of the angle, which is a measure of turn.

The role of language in measurement

The language of comparison

Rather than using units of measure, much of our early measures work involves lots of direct engagement with items and learning how to make a direct comparison of how heavy they feel, their length by lining them up against each other, and so on. Here we would be seeking to teach vocabulary such as 'taller' (where only two items are being compared) and 'tallest' (for comparisons of three or more items). As well as being aware of the linguistic distinction between the 'er' and 'est' words, we also need to support children in understanding that we are using comparative language. As we see in Figure 7.4 the giraffe is taller than the little boy, but he is taller than the ant, thus the boy is both taller and shorter!

So can we use words like 'tall' or 'heavy' without needing to worry about comparison? If someone says 'goodness, that's a tall tree', it is implicit rather than explicit that it has been compared with other trees and in comparison is particularly tall. It would be wrong, however, to imagine random items could be labelled as being tall or long or whatever. They may be taller or longer than each other or other things, but we cannot pick something and just say 'it's tall' unless in reference to something else. Sadly, I see lots of worksheets which make a poor attempt to tackle this area, so be on the lookout for them and avoid using them in your own teaching.

Figure 7.4 Is the boy tall or short?

Incidentally, do not make the assumption that direct comparison is something we grow out of. As an adult you are still likely to use it on a regular basis. For example, if I want to post something and I am hunting for an envelope, I would not reach for a ruler; instead I would just hold the item up against the envelopes I have and it will be pretty obvious which one it will fit in. If I have leftovers after a meal and I want to save them, finding a suitable container is done by eye. Think about the occasions in life when you use direct comparison and manage to avoid formal measuring. This is about making our teaching as realistic as possible, about being cognisant of the ways in which we use measurement in our lives. We can then provide examples of why we might want to know which is the longest toy, or the heaviest piece of fruit. Themes, such as the Olympics, or topics such as 'Pets' lend themselves particularly well to learning about the application of different forms of measurement.

The language of length

As we have seen above, comparative language needs to be handled quite carefully as there is plenty of potential for misuse. Another aspect to be aware of is the fact that we use so many different terms to denote length. We referred to the giraffe as being tall in relation to the boy, but I recall a student teacher mixing the language 'long' and 'tall' rather indiscriminately, resulting in a poor language model for the children, with examples like 'tall' snakes measured from tongue to tip.

Although 'tall' and 'long' have generally distinct uses, they share the same opposite term 'short'; thus short is used to denote lack of height as well as lack of length. 'Depth' is similarly variable, the depth of a swimming pool being somewhat different to the depth of a piece of furniture. Thinking about the depth of a swimming pool, our answer may use other related vocabulary; that we like paddling in the 'shallow' end, or diving into the 'deep' end. A trip to buy furniture will draw on height and depth, but probably width too.

What about if I want to measure round something? In mathematics I start to come across specific terminology such as perimeter and circumference for 'round the edge' measuring, and whilst learning about circumferences I will also need 'radius' and perhaps 'diameter', all of which are also measurements of length.

Incidentally, length can also relate to time as well as to distance. The journey from Durham to Chichester is a long one; it takes a long time and it is a long way. In comparison, popping to the supermarket is a short journey in both distance and hopefully time.

So next time you plan a lesson on length, think carefully about how you will model the language and draw children's attention to some of the intricacies. Note that so far I have avoided the language of 'big' and 'small'; generic size words which can also be used to compare items, with all the same attendant notes of caution sounded above. This is not to say, however, that we should not use words such as 'big', 'large', 'small', 'tiny', and so on; questions involving such vocabulary have the potential to generate some really rich dialogue (and quite possibly dissension, leading to a better understanding) precisely because of their ambiguity.

Specific relationships and differences in measures

Area and volume

Having thought about length in its various different guises, we need to consider its relationship with both area and volume. Units like centimetres are sufficient for measuring all different types of length (which we might want to think about as one-dimensional measuring), but if I now want to measure something in two dimensions (sticking out this way and that way), then I need a unit that reflects this and hence we get units such as cm^2. The step up to cm^3 reflects that we are now measuring in three dimensions, and can describe an amount of space being taken up by a solid. Much of the area and volume work at primary level involves rectangles and cuboids and draws on children's ability to calculate using multiplication knowledge. For further detail on these and other topics, see Suggate *et al.* (2010).

The link between cm and cm^2 can perhaps most profitably be explored in work combining perimeter and area, noting that children do seem to muddle the two and that there are potential misconceptions.

Common misconceptions

'As perimeter increases so does area'

For many children it feels intuitive that area and perimeter are fixed in relation to each other, and it will be through investigation using resources such as squared paper and geoboards that children will hopefully be disabused of this misconception.

'To work out perimeter I count squares'

If a child gives you an answer of 10 (centimetres or whatever) when working out the perimeter in Figure 7.5, it's likely that they have counted the whole squares (the ones round the edge shown in grey) rather than the edges (illustrated using the arrows) which will give us the right answer: a perimeter of 14 units. Using play equipment, such as fence sections from a farmyard set, is great for modelling the idea of perimeter as an 'edge' measurement, and contexts such as the layout of allotment plots can help to bring the topic to life.

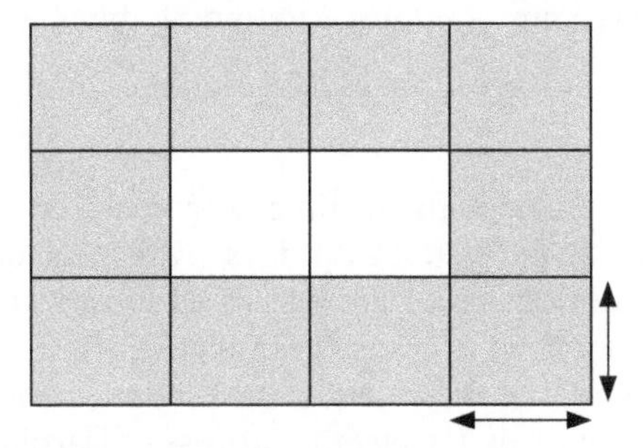

Figure 7.5 Perimeter illustration

Volume versus capacity

Whilst volume and capacity are clearly related (1 cm^3 of water is equivalent to 1 ml of water), it is capacity that is more accessible to younger primary-age children, focusing on how much something holds (the capacity of the item) and gradually getting to grips with terminology such as 'litre'. How much space something takes up usually implies volume and uses the cubic units discussed in the previous section. One possible way of exemplifying the difference between capacity and volume is to imagine a child's lunch box with a plastic flask in it. There is not much room left for sandwiches, fruit, and so on as the flask takes up quite a lot of space (and that space is the volume of the flask). If we then turn our attention to what the flask can hold (a certain amount of squash or soup or whatever), then we are talking about its capacity. Note that this is less than the volume and would generally be measured using different units.

I need to sound one word of caution here, however. Whilst capacity is generally accepted as being measured in units such as litres, in reality it is not always liquid in nature and there will often be gaps between the contents. If I buy a large box of cereal, for example, it has a greater capacity to hold its contents than a smaller box does; mine says it contains 750 g, and whilst they could clearly have fitted a lot more in had they squashed the contents down a bit, I am glad they did not! You may well wonder why I am waffling on about my breakfast cereal in a book about essential primary mathematics, but this 'nonsense' helps to illustrate the messy reality of what measures are all about and could transform your teaching. Good measures lessons are not about learning measuring skills in isolation, but about the relationship between the mathematics and everyday life.

Weight versus mass

When children and adults talk about how much something weighs, what they are actually doing is thinking about the mass of the object or how much matter the item contains, measured in units such as kilograms. Weight is affected by gravity, thus if we were to ever travel to the moon, we could experience relative weightlessness, whereas our mass would remain the same. Whilst as professionals we ought to know the difference in order to be able to engage accurately in dialogue with children, I would not want us to get too concerned about correct usage. We are teaching mathematics to children who are growing up in the real world and people will inevitably continue to use the language of 'weighing' as opposed to finding the mass of things.

Time

In the early stages, rather than telling the time, what children need to grasp is that time is parcelled into particular packages: into days (which include mornings and afternoons and evenings and night-time) and weeks and years and so on. Once children grasp the basics, they then need to become familiar with the smaller subdivisions: that days contain 24 hours and that these can be broken into minutes (and later seconds). Only then will children be ready to start the process of learning to tell the time, which itself involves getting to grips with two different types of clock: analogue and digital.

As children will come across both types of clock in their everyday lives, I suspect that, for many, learning about them in tandem will be the most appropriate approach. The children hopefully then begin to recognise simple relationships, such as the o'clock analogue times versus their digital counterparts. On the hour I find it helps to stress the fact that it is however many hours and no minutes, which is perhaps more obvious on a digital clock. Analogue clocks employ a sort of double numbering system, where the numbers we see on a regular clock face denote the hours, but the same position also denotes a certain number of minutes (and seconds too). If you have ever stopped to analyse what is involved in learning to tell the time, then it will come as no surprise that it takes some children a while to master it! Recognising the direction in which the hands are travelling is crucial; if children have not grasped this directionality, they will struggle to comprehend 'past' and 'to' times. Draw their attention in particular to the progress of the hour hand: note when it has just gone past an hour, when it has almost reached an hour or is about half-way between the two. If you are not able to use a real clock, then a geared analogue clock (one of the plastic education ones with cogs) will be your best bet as this will ensure the hands are in the right places. Use opportunities presented by the children to discuss the fact that we can say something special about 15 minutes past or to the hour.

Common misconception

'The time one minute after 3:59 will be 3:60'

As time is parcelled into sixties, children may find it challenging to remember that the time one minute after 3:59 will of course be 4:00.

As well as learning to identify particular moments in time, and to understand that they can be expressed in different ways (as I write this it's just gone quarter past five; 5:17p.m. to be precise), children also have to grapple with lengths of time: the duration of a particular event or the time which elapses between events.

Activities for measurement teaching

Some of the following activities target a single area of measures, whereas others can be altered to focus on a particular area, or used so that the children work across several areas of measurement at the same time. When planning to use these activities, consider which part of your lessons they will prove most useful in, and remember that they can easily be adapted to meet slightly different objectives for different age groups, for example, by changing the values involved.

Pupil activity 7.1

The ambiguity of 'big' and 'small'
Y1 to Y6

Learning objective: To develop a better sense of the different things 'big' and 'small' mean and to begin to find ways of measuring awkward items.

Having a couple of really oddly shaped items and asking the children to compare how big they are has the potential to generate some really rich discussion which may delve into height, width, mass, and so on. This could involve shapes modelled out of play dough which is particularly nice as they can then be remoulded for easier comparison of the actual amount of modelling material. After a period of heavy rain, looking at different size puddles could be a super starting point for this type of activity; really good mathematical discussion would hopefully proceed beyond surface area to considerations of depth too. This activity could later be extended to focus on estimates of area by superimposing grids of squares on top.

Pupil activity 7.2

Ordering by direct comparison
Y1 to Y2

Learning objective: To understand and use comparative language accurately.

Children holding items and sorting themselves into order according to a given criterion is a simple but very effective activity. The teacher can model the sort of vocabulary associated with measurement, for example 'How tall is your bear?', 'Is it taller than Zac's bear?' and 'Whose bear is the tallest out of everybody's?'. Note that children may need to be prompted to line the bears up, with the heads or paws level, in order to get an accurate comparison.

The ordering could relate equally well to capacity or mass: thinking about containers that might hold the most, or comparing items to discover which are heavier or lighter than others. Incidentally, it can seem easier to feel differences in weight if the items are put into carrier bags and held one in each hand as opposed to holding the actual items.

Pupil activity 7.3

Ordering of measurement values and terms
Y1 to Y6

Learning objective: Ability to identify the order in which measurement information would occur.

In addition to the ordering of actual items, children can also focus on values presented in different ways such as the mixture of millimetres, centimetres and metres shown here. This can be information on individual cards or mini whiteboards, or even on product packaging. Rather than individual activity, I envisage this would be done collaboratively so that the children benefit from the dialogue with each other – arguing if necessary about what goes where! Note the similarity of the values used for my example here; more reasoning is required precisely because of the numbers chosen. It is also important to recognise that underpinning this would be the ability to multiply and divide by 10, 100, etc., using secure place value knowledge.

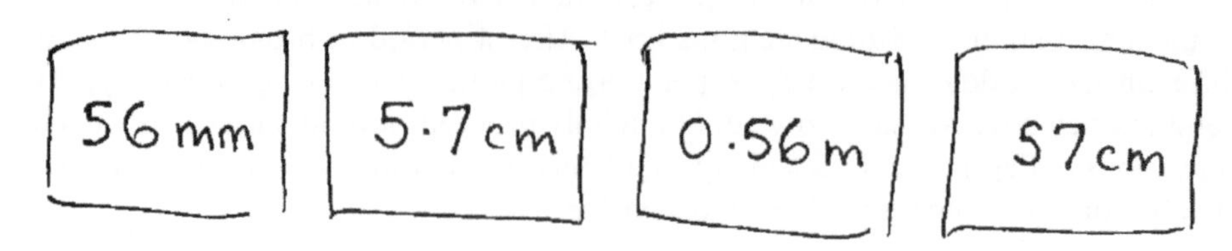

As with the other activities, the focus could equally be on mass, capacity, and so on. Lengths of time are also fun as these can be given in a mix of seconds, minutes and hours, and you might even include some ambiguous ones such as 'morning'! With younger children pages from an old calendar could be used for ordering the months of the year, and of course opportunities to practise ordering the days of the week are needed too.

Working across more than one group of units makes the ordering more of a challenge. With a mix of metric and imperial measures children then need to employ knowledge of equivalent values to complete the ordering. For example:

- centimetres, metres, inches, feet;
- grams, kilograms, ounces;
- millilitres, litres, pints.

Pupil activity 7.4

Sorting measurements
Y3 to Y6

Learning objective: Ability to distinguish between different measurements in relation to real-life contexts.

Temperature can be explored in various different ways, and for older children climate charts provide a wealth of information, for example, to help people decide what to pack for their holiday (see linked activity 8.9).

Sorting activities also work well. Given the following scenarios and temperatures, which do you think would be 'just about right', 'a bit too warm' or 'a bit too cold'?

- Milk in the fridge, 4°C
- Roasting a chicken, 100°C
- Sunbathing, 12°C

The sorting could take a more active approach were these three labels put on three different walls in the classroom, with the children voting by turning towards or pointing at their choice. We might then need to build in some opportunities for the children to investigate answers they are not sure of, such as asking at home what temperature you might cook a chicken at. This could be adapted to other aspects of measures – sorting statements as to whether they would be 'a bit too heavy' or 'a bit too short' or whatever.

Pupil activity 7.5

Area and perimeter relationships
Y4 to Y6

Learning objective: To understand that area and perimeter are not fixed in relation to each other and to learn ways of calculating them.

Have children investigate the possible rectangles that could be generated out of a given number of square units, with numbers like 12 giving several possibilities because it is a number with several factors. Use the resulting rectangles to explore what the different perimeters would be, checking there are no misconceptions regarding what to count.

Once secure in the fact that area and perimeter are not fixed in relation to each other (and not muddling the two), children can be shown a variety of different sized rectangles made up of squares: can they find a quick way of working out the areas? What we want to promote are the links with multiplication; recognising the fact that there are so many rows and so many columns (akin to an array) moves children from counting unit squares towards a more formulaic approach, where the area of any rectangle can be found by multiplying two lengths.

Open-ended tasks have several advantages as advocated in Chapter 1, and here we might want to use just such a task. This one assumes children's familiarity with the formula, and relies heavily on children's factor knowledge, although reverting to use of squared paper could simplify the activity. Give a choice of different area values (ensuring you give some thought to the range) for children to investigate the different rectangles these areas could describe.

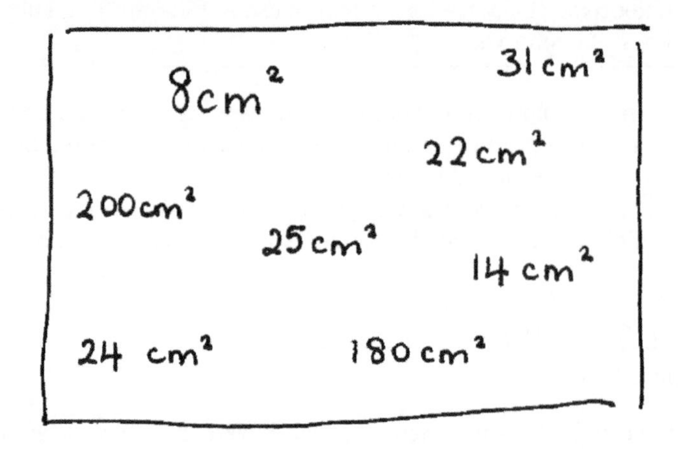

Note that 24 cm² and 25 cm² alone would give an interesting contrast. The former offers a huge number of possibilities: the rectangle could be 3 cm × 8 cm, 6 cm × 4 cm, and so on. This contrasts with the latter which is a larger number but more limited (25 cm × 1 cm or 5 cm × 5 cm) but does alert children to the possibility of square rectangles, examples of which could be used in both activities. I included 31cm² as it is a prime number and there will therefore be only one solution. I put 180 cm² and 200 cm² in because children often love to give themselves the challenge of working with larger numbers.

Either activity could be expanded to look at non-rectangular arrangements (sometimes referred to as 'rectilinear' shapes) seeking minimum and maximum perimeters for fixed areas, or calculating areas by breaking the shapes up or enclosing them in rectangles.

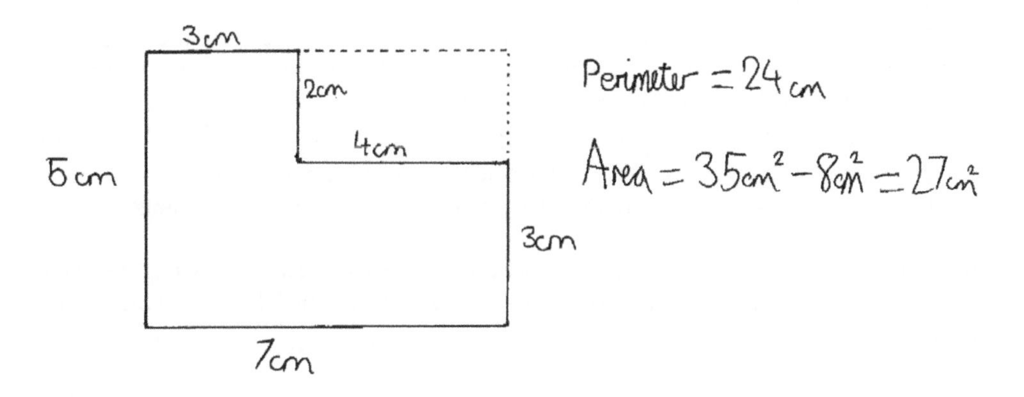

Pupil activity 7.6

Rehearsal of measuring skills
Y1 to Y6

Learning objective: To develop the sorts of skills associated with accurate measuring.

Early measuring skills are mostly learned by having a go at finding out how long or heavy something is, or how much it holds, but adult guidance is needed to draw the children's attention to the accurate use of the apparatus, whether a tool such as a ruler, or units such as Multilink.

A nice way of giving the children ownership of the skills is to work together to produce booklets or posters describing in words and pictures how to measure particular attributes. Note the possible areas of difficulty (suggestions in brackets) as these are what you want the children to realise and make explicit for their peers.

- If you want to find out the length of something... (Make sure you line up your Multilink from the start of the item and you do not leave gaps; make sure you look to see where zero is on the ruler and start there; make sure you have your metre stick the right way round.)
- If you want to measure round something... (Make sure you are looking at the right end/side of the tape measure; make sure your string goes round all the wiggles of what you want to measure.)
- If you want to find out how heavy something is... (Make sure the balance is even before you put anything in; make sure you count the Multilink or add up the masses carefully; make sure you zero the kitchen or bathroom scales before you begin.)
- If you want to find out how much something holds... (Make sure you get down to the same level as the jug before you try to read the scale; think about why you might get different sorts of answers depending on what you want the container to hold.)

- If you want to find out how big an angle is... (Make sure you have the protractor the right way up and are reading from the right end; think about expected answers, for example 'this is a bit bigger than a right angle'.)
- If you want to find out how long something takes... (Think about why you might choose to use a certain type of timer, such as a sand timer; make sure you know how the buttons on your timer work so you do not lose seconds fiddling!)

In addition to the ideas above, the older children will also need to rehearse the skills associated with finding out what values appear in between the labelled values on any measurement scale.

Pupil activity 7.7

Large but not necessarily heavy
Y1 to Y4

Learning objective: To appreciate that mass varies according to what an object is made of rather than its size.

Common misconception

'This is big so it must be heavy'

One of the earliest misconceptions associated with mass is that large items will automatically be heavy. Whilst we know that this is not necessarily the case, children will need opportunities to experience this for themselves. The worst thing we could do would be to unwittingly reinforce this stereotype by providing large heavy items and small, lightweight ones.

A nice activity is to collect a range of different balls of varying sizes and masses to play with; throwing them, catching them, rolling them will help children to appreciate that they feel quite different.

Alternatively, you could prepare some huge lightweight boxes, versus small boxes filled with heavy items, and allow the children to play with the items in the context of helping a story character move their things.

A third option is to fill a suitcase with a mixture of different holiday items to investigate and discuss. All of these activities have the same aim: to challenge children's early assumptions.

With older pupils it is possible to explore a related theme by investigating costs per kilogram (or 100 g or other unit). Comparing different types of fruit and vegetable, they will find that the heaviest choices do not necessarily mean the most expensive.

Pupil activity 7.8

Varying capacities
Y1 to Y3

Learning objective: To appreciate that capacities can be the same whilst appearing different and to develop skills associated with measuring capacity.

As with other areas of measures, we would expect much capacity work to involve practical activity, and use of sand and water trays is ideal for exploration of non-standard measures such as cupfuls, as well as work in millilitres. A suggested progression for the 'play' is as follows:

- Investigation of whether bottles hold more or less than each other by pouring carefully from one into the other.
- Comparison of how much each container holds by counting the cupfuls that fit into each bottle.
- Knowing the capacity of the cup in millilitres or using a measuring jug to identify the capacity of each bottle.

Common misconception

'This is tall so it must hold a lot'

The greater the variety in container shape the better, as one of the things children will need to come to appreciate is that containers may look very different and yet hold the same amount. The examples in the photograph are all half litre bottles, but they vary in height and shape.

Pupil activity 7.9

Reading scales
Y2 to Y6

Learning objective: Ability to understand measurement scales.

Having whole classes involved in practical activity at the same time would be likely to be a logistical nightmare and quite possibly you would not have enough equipment anyway. Associated activities can therefore be a blessing. Combining work on fractions (in particular, quarters) with that on capacity, this child was given bottles with different totals and worked out how many millilitres the bottles would hold at different points. Note that it is easy to vary this activity according to how much the bottle holds, with something like 500 ml being more challenging for quarters than, say, 800 ml. Many of you will be familiar with interactive teaching programs available online, use of which could support an activity like this one.

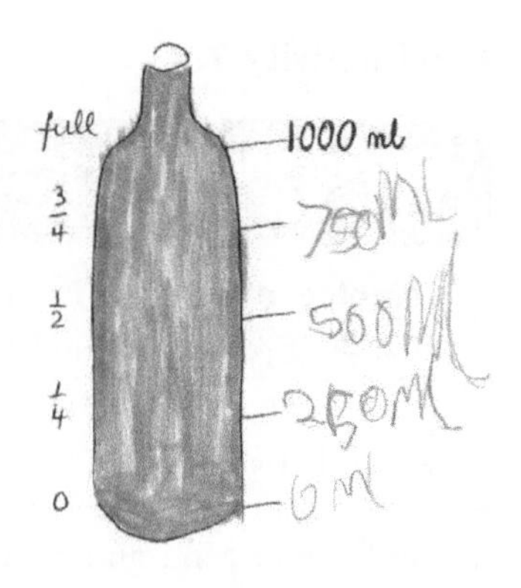

The activity need not have a capacity or fractions focus of course. We might choose to concentrate on the intervals on a ruler or a tape measure, identify numbers of degrees in the gaps on protractors or thermometers, compare kitchen and bathroom scales, and so on. Allowing children some freedom to draw their own pictures and decide the points to identify can remove the need to make differentiated worksheets.

Pupil activity 7.10

Estimation
Y1 to Y6

Learning objective: To develop the ability to estimate with increasing accuracy.

A fun way of practising and improving estimation of length in metres is to fix a starting line in the playground or hall, and for the children to work in small groups to throw a beanbag from the line. How many metres from the line do they think it landed? Upon checking, compare the estimate with the actual distance and have a good look at the gap between the starting line and the beanbag – such attention to detail will help children become better at estimation over time. With each successive attempt any adult supporting the activity can help to model appropriate vocabulary regarding the beanbag's position – for example, 'I think it's a little bit further away this time.'

Estimation, or 'best guessing', is always ideally based on this cycle of estimating, checking for accuracy, repeating the activity discussing estimates in relation to the first attempt, and finally checking again to see whether the guessing has improved. Dialogue taking place is crucial here, otherwise there is a danger that children replicate the same sorts of estimating errors repeatedly; it is also valuable to familiarise children with something of the appropriate unit length (or whatever) before starting the activity – for example, 'do you remember when we used these metre sticks…?'. Estimation could of course involve mass, time, capacity, or angle, for example.

Pupil activity 7.11

What can be measured?
Y4 to Y6

Learning objective: To develop a greater appreciation of the units associated with different areas of measurement.

One good way of getting children to think carefully about different ways in which things can be measured and the associated units is to present the children with a range of values without units; just a small number is ideal, as the activity already requires children to think about a couple of different things. Describe a scenario such as walking from school to the local park. Some children might suggest 500 m, and others 10 minutes, with both answers being judged to be reasonable. I would probably have the children working in pairs to encourage some joint decision-making as to suitable answers. Their answers will allow you to assess their reasoning, as well as giving some indication of which forms of measurement children find it difficult to estimate with. Other scenarios could be:

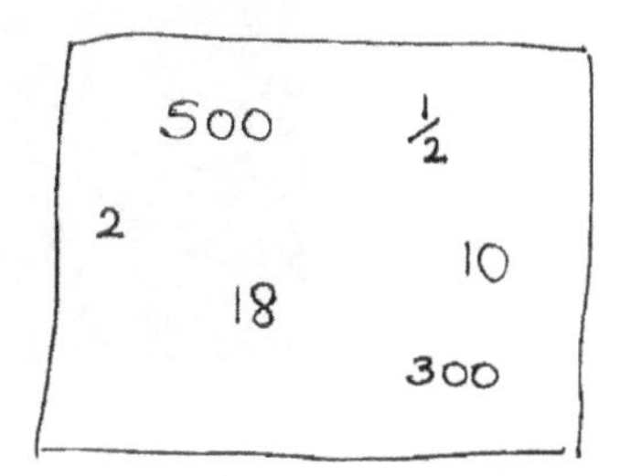

- The width of my little finger nail
- The size of an apple (purposefully ambiguous!)
- The weight of my suitcase
- A winter's day
- How much my mug holds

Check out anything you are unsure of (how much does an average mug hold?) and try to think of some other scenarios you could add to the list.

Pupil activity 7.12

Mass information on packaging
Y3 to Y6

Learning objective: To understand terms relating to mass and develop associated estimation skills.

A fun way of exploring how heavy a certain number of grams feels is to explore packets of different foodstuffs, some of which are full, others almost empty, and so on. The children can be asked to investigate and compare the results they get using scales versus the information given on each packet, and why there might be a difference even if a packet is full. This often opens up some interesting discussion about packaging materials or about the information given on the packets, such as symbols, drained weights and the terms 'net' and 'gross'.

Pupil activity 7.13

Mass by balance
Y1 to Y4

Learning objective: To have a secure understanding of how a bucket balance scale works.

Much early work takes place using balances: scales which work much like a seesaw. In fact this may be a useful analogy (or better still, build in a trip to a local playground) to help children get to grips with what is happening. If children are seeking to compare two items by placing one on either side, understanding that the heavier side will go down is vital. To then find out the mass of an item (or

how much it weighs) we need to balance the item against non-standard units ('My book balances 13 conkers'), to use something like Multilink which is semi-standard, or to use masses labelled with standard units. By this stage children's mental calculation skills need to be good enough to add up masses of different values.

I often build in a context, such as finding out how heavy parcels are in order to work out how much it would cost to post them.

Pupil activity 7.14

Telling the time
Y1 to Y4

Learning objective: Ability to read increasingly complex times on both types of clock.

When starting out it is worth watching just the progress of the hour hand since we can pretty much tell the time from its position. Whilst this could be the focus of a lesson, it is also an ideal activity for those spare moments, such as being lined up and waiting to go to assembly. Having begun to appreciate what the progress of the hour hand tells us, we can relate this to the display on a digital clock, where it is the number of minutes that can tell us that we are just past one hour (4:03) or nearly at the next (4:58). Remember the possible misconception here: children may not recognise '58' as being close to the next hour. I find humour a great teaching approach; thus joking around with 4:98 and other impossible times (sorting them from the possible ones) would be a good way of making the teaching point and addressing misconceptions.

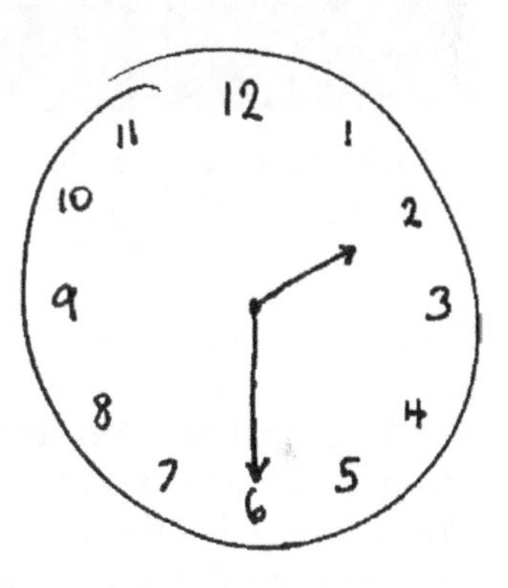

Another type of 'impossible time' is the one shown here, which provides a wonderful starting point for mathematical discussion and will help illustrate how to read the time on a proper analogue clock. What do the children think the time might perhaps have been? Half past two? Two o'clock? Half past one? How would these be written as digital times? Where would the hands need to be, to be displayed properly?

Pupil activity 7.15

Lengths of time
Y2 to Y6

Learning objective: Ability to calculate the difference between two given times.

TV guides are a useful source for exploring lengths of time, but it could be any scenario where start and finish times are key, such as baking something for a certain length of time, or booking appointments. Thinking about an imaginary excerpt from a TV guide, children could be asked how long a *Strictly Come Dancing* fan (guilty as charged!) needs to set aside to watch the show. They can then go on to think about what they would choose to watch and how long their viewing would last. Note that listings which cross the hour are potentially more challenging, and, just like learning to bridge the tens, some children will need support with this.

Following activities involving simple time schedules such as TV guides, some children will be ready to progress to timetables (in particular, those associated with bus and train travel) which incorporate the 24-hour clock. Leading on from this flight schedules could be explored, where differences between time zones would need to be taken into account!

Pupil activity 7.16

Angle activity
Y1 to Y5

Learning objective: To understand angle as a measure of turn and develop familiarity with degrees as the unit of measurement.

Early angle work is best explored practically and will involve the children physically making full, half and quarter turns both clockwise and anti-clockwise (see activity 11.10). The relationship between quarter turns and right angles can then be explored, eventually leading to the fact that a right angle has a certain number of degrees and is written '90°'. Children often make their own angle eaters (cardboard with 90° sections cut out of them) and go round the classroom looking for right angles; in addition to this, try to retain a sense that angle is also dynamic, enabling us to describe an amount of turn. Robots such as 'Roamer' are useful in this respect.

Pupil activity 7.17

Familiarity across different areas of measurement
Y3 to Y5

Learning objective: To appreciate the application of measurement in a real-life scenario.

Get the children to work in groups to list different things that can be bought in a supermarket where the quantities are expressed in different ways: just by number (for example, melons); by weight/mass (onions); by length (sticky tape); and by capacity (juice). But watch out for some surprises. Looking at packaging together will alert children to products like yoghurt which seems to be sold in grams whereas we might have thought it would be millilitres. Note, too, that it can be combinations of number and the others – a fun activity is to roll a dice to specify the number of the item or packet and work out the overall mass, capacity or length. Costs can also be incorporated.

Recipe books also offer a wealth of examples of measurement in practice, and are a potentially valuable source of investigation. I have recipes jotted down which ask for ingredients such as flour in cupfuls (typically American recipes do this) whereas others come from older UK sources and give imperial amounts, or the recipe might ask for grams – and to be honest I often estimate using a tablespoon, knowing roughly how much this would be.

Pupil activity 7.18

Conversion of measures
Y5 to Y6

Learning objective: To develop strategies to enable conversion between measurements.

Rather than using an algorithmic approach to converting measures, my preference, to start with at least, is to build from a known fact: if I know that an ounce is roughly 25 grams, then I know 4 oz is about 100 g, and so on. Note that an ounce is actually closer to 28 g, so children need to be aware that this is only approximate and would become more inaccurate were larger quantities involved. Consider the different areas of measurement for which children can be given a starting fact and asked to build other facts from it, for example:

- There are about 8 km to every 5 miles, so 10 miles must be 16 km (double), 1 mile must be 1·6 km (÷ 10), 15 miles must be 24 km (adding)
- The exchange rate is currently $1.57 to £1, so...
- There are 12 inches to a foot and about 3 feet to a metre, so...

Graphs are also a useful way of seeing the relationship between two units designed to measure the same thing, especially for something like temperature. Building from a single known fact (0°C is equivalent to 32°F) is not an option here since temperature is not a ratio scale.

Summary of key learning points:

- purposeful measuring activity keeps real-world applications in mind and will involve hands-on opportunities;
- standard units are necessary to be able to compare measurement values with others, but we can also consider direct comparison opportunities;

- a secure understanding of the number system underpins children's ability to work within measures;
- measurements are only ever approximate and different activities require different levels of accuracy;
- the language of measurement should be modelled carefully;
- examples selected should actively challenge misconceptions.

Self-test

Question 1

When measuring angle we are using (a) a ratio scale, (b) an interval scale, (c) an ordinal scale

Question 2

Time is slightly more complicated because it works on two different types of scale: (a) interval and ordinal, (b) ordinal and ratio, (c) ratio and interval

Question 3

If 1 ounce (oz) is approximately 28 g, 100 g is (a) just over 4oz, (b) just under 4 oz, (c) exactly 4 oz

Question 4

A rectangle consisting of 4 square units (a) could only have a perimeter of 8 units, (b) could have a perimeter of 8 or 10 units, (c) could have three or more different perimeter lengths

Self-test answers

Q1: (a) We know that angle is a ratio scale because we can make statements about one angle being however many times bigger or smaller than another.

Q2: (c) Time is both a ratio and an interval scale: ratios are fine when we are comparing lengths of time, but time is an interval scale when thinking about specific moments in time.

Q3: (b) As 4 oz would be 4×28 g = 112 g, 100 g must be just under 4 oz.

Q4: (b) We are limited here by the number of different rectangles we can draw: just two examples with perimeters of 8 and 10 units.

Misconceptions

'12°C is twice as hot as 6°C'

'This has long "arms" so it must be a big angle'

'As perimeter increases so does area'

'To work out perimeter I count squares'

'The time one minute after 3:59 will be 3:60'

'This is big so it must be heavy'

'This is tall so it must hold a lot'

8

Data

About this chapter

Before exploring the particular types of graph typically studied with primary-age children, this chapter will discuss some of the background to data handling to support you in ensuring it is well taught. The content is organised under the following headings:

- Statistics and the data handling cycle
  - Introduction to data and foundation skills for data handling
  - Qualitative versus quantitative data
  - Unpicking the data handling cycle
- Ensuring progression in data handling
  - Making the most of the data handling cycle
  - Using resources to aid progression
  - A secure grasp of the language of data handling
- Specific types of graph
  - Pictograms
  - Block and bar graphs
  - Pie charts
  - Scatter diagrams
  - Line graphs
  - Sorting diagrams

What the teacher and children need to know and understand

Statistics and the data handling cycle

Introduction to data and foundation skills for data handling

Before discussing the teaching of data handling and exploring various different types of graph and chart, it is important to understand a little of the background to this branch of mathematics. The term 'data' refers to facts or information about a chosen topic, hopefully gathered in a reliable manner such that you can be sure of their accuracy. Note that it seems to be accepted practice these days that the word 'data' can be used either in the singular or the plural. You will probably also have come across the word 'statistics' which tends to be used to describe the information once the data has been processed in some way. One rationale for this is that spotting patterns and relationships within the data becomes easier once it is organised into some sort of chart, graph or diagram, and statistical information can include averages and so forth. These and other 'labels' for the ways in which data is presented seem to overlap to a certain extent, so expect to find various references to them here and in other materials.

Statistical literacy, a reasonable ability to understand data (as well as chance, the topic of the next chapter), has become increasingly important as technological advances mean that we have access to huge quantities of data on a regular basis and are bombarded with statistical information as part of our everyday lives. Consider buying a new car or household appliance. You will have access to a wealth of data on each one and, with a good grasp of what it is telling you, you can compare products and hopefully make an informed choice. I chose a new sleeping bag recently and even that involved consideration of data relating to mass and dimensions, as well as temperature ratings for comfortable sleeping and for use in extreme cold! Sadly, however, data is not always presented clearly and can sometimes be quite misleading; it is therefore essential that children develop a secure understanding of this aspect of mathematics.

Ability to identify similarities and differences in order to sort things is a fundamental concept underpinning data handling. Without this ability to discriminate a child will struggle with data handling tasks, such as finding out how many people have hair of a particular colour. Tasks such as this one can be undertaken practically of course; literally getting the children into groups alongside other children with similar coloured hair to illustrate the basis of what graphs and charts will show at a later stage. Young children will benefit from lots of opportunities to sort, group and compare actual items before progressing to recording such activities on paper or on the computer.

Categories are not necessarily straightforward, however; I would like to think I belong in the brown hair group, but in reality I have an ever increasing amount of white hair so, given a simple choice between groups, which colour group would I belong in? We might decide to extend our fact-finding mission to encompass different types of hair, mine being thick and wavy. Learning to deal with categorising things, including tricky information, is therefore a key aspect of data handling.

Having gathered my data into appropriate categories, I can then apply my number skills (such as those explored in Chapter 2) to compare and contrast the information, such as identifying the number of people with hair of a certain colour or type, spotting which groups have the most or fewest people and perhaps that one of the categories is empty. Remember I have only gathered data from a small sample, and children will eventually begin to consider whether data gathered might or might not be representative of a larger proportion of a given population.

The hair colour scenario described above was just an example, but real-life contexts are a vital part of data handling, given that data has to be about something. Other curriculum areas, and even other topics in mathematics, lend themselves to being explored through data handling, so aim to exploit those cross-curricular and real-life opportunities wherever possible. Although the main aim of an activity might be to develop data handling skills, the relationship with another topic is generally mutually beneficial; I understand more about different shapes as a result of sorting them using a sorting diagram, but I also understand more about sorting diagrams. (For more on this, see the later section on sorting diagrams and activity 10.12 in the chapter on shape.)

Whilst much of the data children work with will be 'primary data', data which has been collected first hand, remember that we can also access 'secondary data' as the basis for our data handling lessons. This could include data such as statistics relating to athletes' performance in the Olympics, as we see in activity 8.15. We were not there to collect the data ourselves but can find out information about how far, how fast, and so on. See also activity 8.14, which relates to census data.

Qualitative versus quantitative data

Data exists in two main forms: qualitative data, which is organised into categories such as 'people with brown hair' (sometimes referred to as categorical data); and quantitative data, which is numerical and typically relates to some form of measurement. This can be 'discrete', where the numbers take only certain values, or 'continuous', where all values are possible.

Categorical data is the earliest type that children will work with; for example, finding out how many people have blue eyes, or like curry, or how many of a certain type of vehicle pass along the road outside the school. Once the data has been allocated to a category, the number of entries in that category can be counted. Mathematicians sometimes refer to the 'frequency' to describe the number of times something occurs. When we are trying to count something that is happening quite fast we might choose to use a skill such as tallying which provides a quick and easy mechanism for keeping track of the numbers involved. It also proves useful for recording events repeated many times, such as keeping a note of the numbers of heads and tails when tossing a coin many times. For more on this theme, see Chapter 9.

Common misconception

'This tally tells us that 5 cars went past'

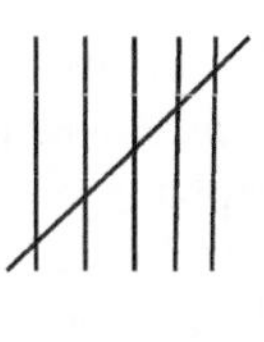

At a quick glance we might agree with the statement, but upon closer inspection you will notice that the picture actually contains six strikes or tallies. Getting used to the fact that the fifth strike is the one that goes diagonally will hopefully come with practice.

To illustrate the difference between the various types of data, let us think about feet and footwear! What are you wearing on your feet at the moment? I'm wearing sandals today, and this is an example of categorical data; if we wanted to make a graph detailing what a group of people are wearing on their feet, the various categories would all be types of footwear (with perhaps an extra category for those of you with bare feet!). Our graph would therefore show qualitative data.

To gather quantitative data we might turn our attention to shoe size (discrete) and foot length (continuous). If measure the length of my foot with a ruler I will obtain a measurement answer. Quite possibly the greatest level of accuracy we could manage with a standard ruler would be to get an answer to the nearest millimetre. Having said that, we should appreciate that measurement is continuous and values do exist in between each millimetre, but we are ignoring them for practical reasons. The measurement scenario and the links with the continuous nature of the number system help to convince us that foot length must be an example of continuous data. Another way of checking is to consider a larger data set. If a room full of people measured their feet, we would end up with a table of many different results occurring along a continuum from the person with the shortest feet to the person with the longest feet. The fact that the results lie along a continuum where all answers are possible, and we are likely to get many different answers, helps to convince us that the data generated is continuous data.

Shoe size, in contrast, is dictated not by accurate measurements of your feet to the nearest millimetre, but by the fact that shoes are only sold in whole and half sizes, or in European sizes, with both systems using numbers to categorise the differences in size from one group to the next (I wear size 4, equivalent to European size 37). Neither system is a continuum as such, since we do not get shoe sizes that fit in between the sizes on offer; there is no point in my saying in the shoe shop that my feet are a bit too large for a $3\frac{1}{2}$ but a bit too small for a 4, so can I have a pair of size $3\frac{3}{4}$? The discrete nature of the shoe sizes, with no sizes in between, gives a clear indication that shoe sizes are discrete data. The other clue once again comes from considering a larger data set. If this time we ask our room full of people to tell us what size shoes they wear, the answers will fit neatly into the predetermined sizes. We can count the numbers of people per group (the frequency) and make statements about how many people wear a certain size shoe; this also helps to persuade me that shoe size is an example of discrete data. Thus some quantitative data can be counted in much the same way as categorical data entries, such as the number of children achieving a particular mark in a test: four children scored full marks, two children got 5 out of 10, and so on.

Whilst children can work with continuous data from quite a young age, the teacher needs to be aware that such data often needs to be presented differently in graphs in comparison to discrete data or categorical data. As the children get older, they should gradually assume greater responsibility for decisions about the ways in which data is presented, distinguishing carefully between the various types.

Note that whilst both shoe size and length of foot are examples of ordered data sets, this is not the case for much categorical data, such as information about types of shoe. There would be no rationale for prescribing an order as to which type of shoe should be listed first, last or in any particular position in between. Graphs showing something like favourite crisp flavour would be exactly the same; the order chosen could only ever be arbitrary, and even if we decided to record them in alphabetical order, this choice would not relate to the data in any meaningful way. In contrast, if we made a graph to present the data about shoe size we might consider it rather odd if the sizes were not shown in order. Categorical data does occasionally have a relevant order, however; for a graph with information about the month in which people were born we could reasonably expect the months of the year to be presented in order.

Unpicking the data handling cycle

Data handling is often described in terms of a cyclic process, and this idea of a loop is pertinent whoever is researching a topic for which some data would be useful. The idea is that you start out wanting to know something and the data handling process hopefully helps you to find it out. Data handling therefore gives children a potential opportunity to engage in problem solving for real.

From a school perspective, not every learner will go through the whole process every time, although taking up those cross-curricular opportunities may mean that data can be collected in a different lesson before being processed in maths. There will certainly be lessons focusing on just one element of the cycle, but over a period of time children of all ages should experience all aspects of the data handling cycle:

- *Questioning*. Someone embarking on a purposeful data handling activity will typically have a question they wish to answer or something they want to find out more about.
- *Planning*. Before actually gathering any data, some planning will ideally take place to decide what information is needed and how it will be recorded.
- *Gathering*. The data is collected. This could be thought of in terms of fieldwork.
- *Analysing*. Analysis of the data requires taking a really careful look at the information that has been collected (often referred to as the 'raw data') and involves skills such as counting and comparing. In basic terms it is about sorting the data out.

- *Presenting*. Once analysed, it can be decided how to present the data to make the information as clear and accessible as possible, especially if it is to be shared with an audience. This could involve deciding to present numerical data in a table or creating some sort of graph or chart.
- *Interpreting*. Interpretation of the data revisits the idea of looking very carefully at the information, now presented in a more organised format. Suitable conclusions can hopefully be drawn as a result.
- *Answering*. This is about closing the loop; the original question has been answered or more information gleaned about the chosen topic. What sometimes happens in reality is that the data handling activity may in fact have given rise to further questions, so it may be necessary to go through the process at least once more.

Given the realities of planning for a class, the question or focus of the investigation may well be specified by the teacher, steering the younger children in particular. Try to select a topic or question that will interest the children who will investigate it. Any of you who have undertaken research as adults will know the importance of this since levels of interest have to be maintained during what can be quite a lengthy process! The richer the original question, the better, since low-level questions tend to negate the graphing stage. Check out the activities you want to use; if questions asked can be easily answered straight from the data collected (perhaps presented in a table), then the construction of a graph may lack purpose.

In order to ensure that all children experience the data handling cycle in full, schools may wish to identify at least one topic per year group that would lend itself to asking and answering a question using data handling techniques. Clearly the topics and the ways in which they are approached would need to reflect appropriate advancement of data handling skills from one year to the next and would gradually incorporate different types of graph.

Ensuring progression in data handling

Making the most of the data handling cycle

One of the main features of progression with regard to data handling is that children gradually take greater responsibility for gathering and dealing with full sets of data as opposed to just contributing their own information. Take, for example, the popular 'Walk to School' weeks; I suspect numerous children are involved in data handling during those weeks as the schools gather data about how many people have walked to school. The data is purposeful; it is to find out how effective the initiative is at getting children (and parents) out of their cars and walking to school, with all the attendant health benefits as well as ensuring there are fewer cars on the road, reduced parking congestion, and so on. But on such an occasion the child typically only needs to know how they got to school and there is therefore a danger that they may be dealing with just one tiny bit of data, and this can result in very low-level data handling tasks. To strengthen such activities teachers need to consider how the focus on travel to school can be utilised to engage children with the whole of the data handling cycle at an age-appropriate level. This may mean designing data collection sheets to use with younger classes, grappling with large data sets, deciding how to present the information clearly using graphs or charts, and suggesting conclusions that could be drawn from the data. Quite possibly children also need to consider what transport arrangements exist in a 'normal' week in contrast.

Another aspect of progression relates to how involved children are with the messy realities that can be encountered, starting at the planning stage of real data handling cycles. Imagine we wanted to find out about people's favourite sandwich fillings. Would you try to set up some categories in advance so that when you start asking people, you have boxes ready to tick? Or might this limit people's choices and prevent you getting the desired information you are really after? Decisions at the planning stage may well be pragmatic and should depend upon the purpose of your research.

If the intention is to identify which types of sandwich would be suitable for a children's party, you might reasonably decide to list the fillings you would be willing to offer and use your research to check which choices would be most popular. The data gathered might result in making lots of ham and cheese spread sandwiches but avoiding making any with jam since your research suggests no one will eat them. Of course this is not strictly true; if you only ask the children for their favourite choice they will only give you one answer. Perhaps jam is every child's second choice and therefore when eating several sandwiches at the party they might choose to eat the jam ones. So perhaps gathering data about least favourite sandwich fillings would also have been useful? What else might you suggest?

Progression in data handling is assured if you are giving children opportunities to grapple with the sorts of issues we are exploring here, but so far we have only considered use of predetermined categories. If you actually want to know people's favourite fillings you might have to allow entirely open-ended answers in order to establish someone's preference for cheese and sardines or whatever. This can make the analysing and presenting stages of the data handling process much harder, and sometimes 'messy' qualitative data is dealt with by grouping all the different answers into broader categories. Even this has its problems, though; I thought I might have a category for fishy sandwich fillings where I could put the tuna and fish paste lovers, along with the person whose favourite sandwich contained cheese and sardines. But if I also had a category for dairy choices, such as cheese, the combination sandwich would belong in both categories. Whilst there are types of chart which deal with category overlap, the aim with much data handling is to ensure that categories are discrete and do not overlap in order to avoid any confusion arising from double entries. I have to be honest and admit I am using awkward examples on purpose; as teachers we have to appreciate some of the difficulties that can arise with real data handling, though not in order to avoid them with the children. It is far better that these difficulties do arise; they can then be openly discussed and the children's understanding furthered as a result.

Knowing that you have asked everybody once and only once is a key skill in data handling if you need to ensure that everyone's opinion will be reflected in the final data set. At first children may find it difficult to keep track of who they have asked, and you might decide to give children a list of names to help them out in the early stages. In the long term this is a skill the children need to develop, however, and they will only do so if they have a chance to have a go, even if they make mistakes along the way.

Common misconception

'There are 27 people in my class and I have asked everyone what their favourite fruit is: 10 children chose apples, 10 children chose strawberries and 10 children chose bananas'

The fact that the figures add up to 30 rather than 27 suggests that at least three children have been asked more than once; it may have been more than three children whose answers were duplicated, with other children missed out altogether, and of course those asked may or may not have given the same answer each time!

Asking a slightly more open-ended question about fruits liked by the children makes this just subtly more difficult with regard to checking that everyone's likes are taken into account; I like all three of the fruits suggested here, whereas some people will like fewer and could potentially not like any of them.

One of the most important things to learn with regard to data handling is that when we get to the graphing stage, not all graphs will be suitable on every occasion. But equally, there are generally alternative ways in which the same information can be appropriately presented. The many different types of graph suitable for the primary age range are explored more fully in the final section of this chapter and through the activities.

Having decided upon the most suitable style of graph, even the interpretation stage can be enhanced to maximise learning. The graph may help us to answer basic questions, such as the favourite sandwich fillings already discussed, but also has the potential to aid prediction or to give rise to new questions. Would the sandwich choices vary if the respondents were older or younger? Do we think the weather makes any difference to people's choices? Do girls' and boys' preferences differ in any way? When presented with data that someone else has collected, children will benefit from being encouraged to ask lots of questions at the interpretation stage. For example, can we be sure that the measurements were taken carefully? Do we know how many people were asked, and were those people representative of a good cross-section of the population? Good mathematicians are not people who take data and statistics at face value!

Using resources to aid progression

For me, the first and foremost consideration when selecting materials for data handling activities is whether they will capture the interest of my current class. Finding a topic that will engage the children is a good starting point. I then have to consider whether the topic will prove a good vehicle for demonstrating the data handling concepts I want to promote. I may have to change my idea if the data generated will not be quite right for a particular type of graph that I want the children to learn about, or where I feel the data would be too simplistic for the age and ability of the class. Getting the focus right takes time, but is important if you want progression to be assured. Note that motivation may be higher if children sometimes have opportunities to collect personal data in some sort of investigation such as who can blow a table tennis ball the furthest in a single puff! This might be part of a science investigation into lung capacity yet would be good fun and incorporate both measures and data handling.

One final but important consideration with regard to topic choice is to check that it is not going to prove a sensitive topic; for this reason children's weight is typically avoided. You will know your class and know what might cause upset, for example, choosing not to gather data about pets when somebody's rabbit has just died.

Having found a topic that will serve your purpose well, one of the simplest resources which will be beneficial is a supply of clipboards. Rather than sitting at classroom tables, gathering data often involves getting up and moving about, and these will help to make collection of the data that little bit easier. Access to paper and pencils is something I take for granted in a classroom but perhaps should not; hopefully your class know where they can find paper (or a small whiteboard) in order to rough out their ideas before attempting to collect any data.

Another major consideration at the planning stage is whether you envisage the children using ICT as part of the lesson, aiming to have a strong rationale for why or why not. With primary-age children one of the main roles for the computer relates to the professional presentation of graphs and charts, something ICT does particularly well. Producing graphs in this way is also quicker than by hand, allowing more time for the interpretation of the material. Because data entered into a computer package can typically be displayed in different graphs and charts at the click of a button, this makes it a powerful teaching tool. The benefits, however, are only fully realised when the practitioner probes children's understanding of which graphs are the most appropriate or clearest way of displaying the data in question. A computer cannot distinguish between different types of data entered into a basic data handling package in the same way a human being can; you may therefore end up with data presented quite inappropriately as a line graph, for example. Note that graphing packages do not always get it right either. For example, you might want to produce a graph showing the months in which children are born only to discover that the software omits the months where there are no birthdays and reorders the remaining months alphabetically.

Technology removes some of the limitations associated with the drawing of graphs, making it much easier to produce graphs such as pie charts and scatter diagrams. It can also result in being able to deal more easily with particularly large data sets. But using ICT as the sole approach for displaying data would seriously disadvantage the children. To draw a graph yourself you have to understand very clearly how graphs work and are constructed; you will no doubt make some mistakes along the way!

Common misconception

'This graph shows clearly how many people like each crisp flavour'

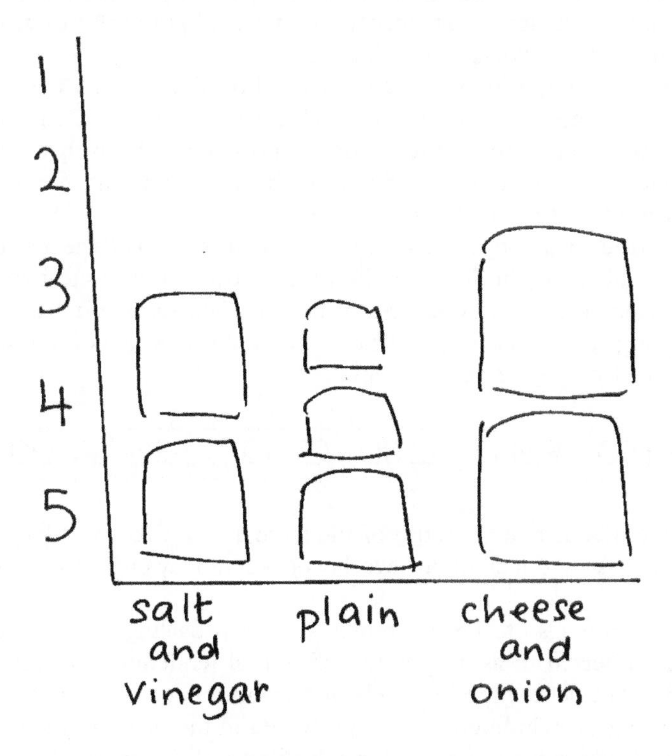

The graph we see here, generated for discussion as a teaching point, exhibits the sorts of errors we do not get if a child only ever produces graphs on a computer, since the computer does not do things like drawing wonky axes or labelling them incorrectly! Note, too, the differences in size between the blocks which would make interpretation more difficult because of the visual confusion. Whilst the computer might help you to avoid such errors, experiencing the difficulties affords the learner greater insight into the purpose and structure of graphs; since computers do lots of the work for the user, these learning opportunities can so easily be lost.

A secure grasp of the language of data handling

As with any mathematical topic, progression in data handling is somewhat dependent on getting to grips with the specific terminology used to talk about data and the way in which it is gathered and displayed. Appropriate use of everyday language is also important; I remember a class collecting data about the months in which their classmates were born telling me that a particular month was 'winning' when of course there was nothing about winning or popularity associated with the data being gathered! Thus everyday words such as 'more', 'most' and 'least' will need to be modelled in context and the children given opportunities to talk about graphs using such words accurately. As well as talking about drawing graphs, children may hear the word 'plot' in relation to graphing, particularly with types of graph showing specific points marked, such as a scatter diagram or line graph.

Many types of graph involve use of 'axes'. For discrete data the horizontal axis often displays the categories and the vertical axis the numbers in each category, but this is by no means set in stone; discrete data is presented effectively in many different ways. When talking about the axes of a graph the horizontal axis might be referred to as the x axis and the vertical one as the y axis.

The statistical term 'range' is used to describe the spread of a set of numerical data and can be calculated by finding the difference between the greatest and least result. Good statisticians retain an awareness of the full data set and are therefore able to comment appropriately. Take, for instance, the following data on the weekly pocket money received by eight 12-year-old children:

£1·50	£1·50	£2·40	£2·80	£3·00	£3·20	£4·00	£20·00

The range here is huge because of the large amount received by just one child; any claims we make about children's weekly pocket money ought therefore to take this into account.

Such caution is also to be recommended when dealing with averages; a statistic sometimes referred to as a 'measure of central tendency', averages should represent the full data set comprehensively and avoid misrepresenting it. When people talk about averages, including averages presented in the press, it is sometimes somewhat unclear which type of average is being discussed. Mean, median and mode are the three main ones introduced in the primary phase, and these are discussed below. Rather than picking any one of the three to work out an average, the intention has to be to select the most appropriate one for the job, the one which will best represent the full data set. Once obtained, this average can then be used to draw conclusions about individual data entries or to make informed comparisons.

Both mean (sometime referred to as the 'arithmetic mean') and median are averages requiring numerical data, whereas mode can be applied to any data set. Finding the mode or modal group is about finding out whether a particular result occurs more times than any other result. Thinking back to the discussion about favourite sandwich fillings, if far more people had fancied ham than any other filling, then 'ham' would be the modal class or group. Note that it is entirely possible to have more than one mode (for example, had a couple of different fillings turned out to be equally popular) or for there to be no modal class at all. The weekly pocket money data happens to have a mode of £1·50, but common sense says that that is just a coincidence and not reflective of the data set as a whole and therefore not a good choice of average in this instance. Mode often works best for large data sets and is particularly useful where mean or median would give nonsensical answers.

The type of average which people seem to associate most readily with mathematics is the arithmetic mean, obtained by adding up all the individual entries in the data set and then dividing by the number of entries. So in the case of the pocket money, all the different amounts add up to £38·40 and if we divide this by 8 (our sample size) then we arrive at a mean of £4·80. This is rather like a division scenario in which we imagine all the pocket money has been pooled and is now being redistributed with the emphasis upon equal shares; the mean tells us what that equal share would be. The disadvantage of the mean with this data set is that we have one rather extreme result which has skewed the average; since £4·80 is more than all except one of the children receive, using such an average in discussion about children's typical weekly pocket money would certainly have the potential to mislead.

The final type of average, the median, will assist us in gaining a more accurate picture of average weekly pocket money, since extreme results do not have quite the same effect upon the median. You will note the entries in the original table were presented in order, and this is important for finding the median, since what we are essentially looking for is the middle value. Had there been an odd number of entries this would have been very straightforward as we would literally have given the middle amount as the median. We have an even number of entries though, and must therefore look in between the two results in the middle of the list. Half-way between £2·80 and £3·00 is £2·90; this is therefore the median.

Note the difference between the mode (£1·50), mean (£4·80) and median (£2·90) for the data set on 12-year-old children's weekly pocket money. Reporting average weekly pocket money honestly therefore requires the reporter to make judicious use of the data so as not to mislead. This has implications for careful teaching about averages; knowing how to calculate each of these types of average is clearly important but teaching children to choose the right one in any given scenario is far more important, albeit a harder task!

Specific types of graph

Pictograms

Pictograms are typically included as part of early data handling software: a type of graph where a picture is used to represent, for example, the pets owned by the children's families. So we might ask everyone in our class about whether they have any pets and, if so, what sort of pet and how many of them. Then, having counted up the results, we could make a pictogram to show the three cats, four dogs, two gerbils, and so on, with suitable pictures chosen to represent the pets in question. All the pictures should be roughly the same size; whilst dogs are much larger than gerbils, what we are representing here is the number of pets, and comparisons between the groups are easier where pictures are a standard size. Note that even pet data can cause data handling controversy: do we count all the fish in someone's pond individually? Whatever the decision, at this stage the pictures would generally represent just one animal each; in time children will be introduced to the idea that pictures can represent larger groups. Most of the examples I have seen over the years start with a picture representing two of something, thus three cat pictures might now represent six cats. Clearly a key would be required to explain this.

Common misconceptions

'As there are three cat pictures in this pictogram this must mean that three people have cats'

Whilst the child making this statement might be right, there are actually two possible misconceptions here. The first relates to the ownership of the pets; we cannot tell from a pictogram whether the cats definitely belong to different people. Maybe it is just one owner who really likes cats! Secondly, the child may have made an assumption that each picture represents just one cat; encourage children out of habit to check whether a pictogram has a key. Incidentally, if we wanted to represent three cats in a pictogram where each picture represented two units, we could end up with $1\frac{1}{2}$ cats pictured!

I would, however, like to recommend that if you are teaching children about pictograms where the pictures represent several of something, that you opt for larger numbers; ideally at least five, but I usually aim for ten or more. This means choosing a data handling context which lends itself to generating large values (see, for example, activity 8.3). This is because it seems to make more sense to the children if having a picture representing quite a few of something saves them the trouble of having to draw lots and lots of little pictures. Look out for pictograms, for example, if you are watching the news, as they are quite often used in the media, though not typically to represent numbers and types of pet! It is not unusual to see something like pictures of money bags with each picture representing government spending of a million pounds. Therefore if we see five pictures for one department's expenditure compared with only two for another we have a quick and easy visual image for the difference in expenditure between the two different budgets. Where large numbers are involved, clearly the figures can be rounded to the nearest suitable number of pictures, although sometimes partial pictures are shown to give a slightly more accurate impression. Note that with pictograms, just as with other types of graph and chart, we might want to draw children's attention to categories which remain empty, such as finding that no one in the class has a hamster.

Block and bar graphs

Block graphs, rather like simple pictograms, allow us to count the numbers represented in each category, as shown in Figure 8.1. This child chose to work on plain rather than squared paper and the blocks vary in size rather. We might therefore wish to draw this child's attention to this feature in other graphs, although the height of the blocks is fairly consistent and they match to the numbering on the left. Theoretically this axis does not require labelling since we could resort to counting the blocks, whereas on the other axis the categories need to be clearly displayed. Just like a pictogram, a block can be used in the longer term to represent more than one of something.

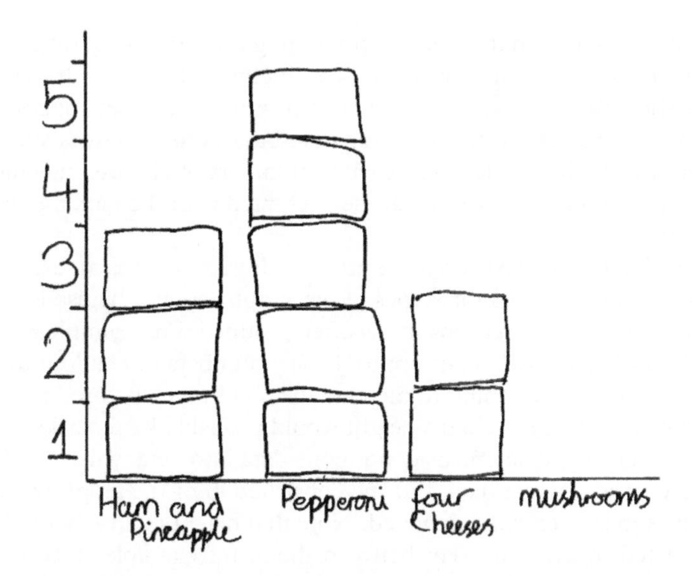

Figure 8.1 Block chart showing favourite pizza toppings

With a bar chart, however, both axes need to be labelled as the bars otherwise have no discernible value. This is illustrated in Figure 8.2 where the same data is shown in a bar chart. Note the contrast between the labelling of the vertical axis from the block chart to the bar graph.

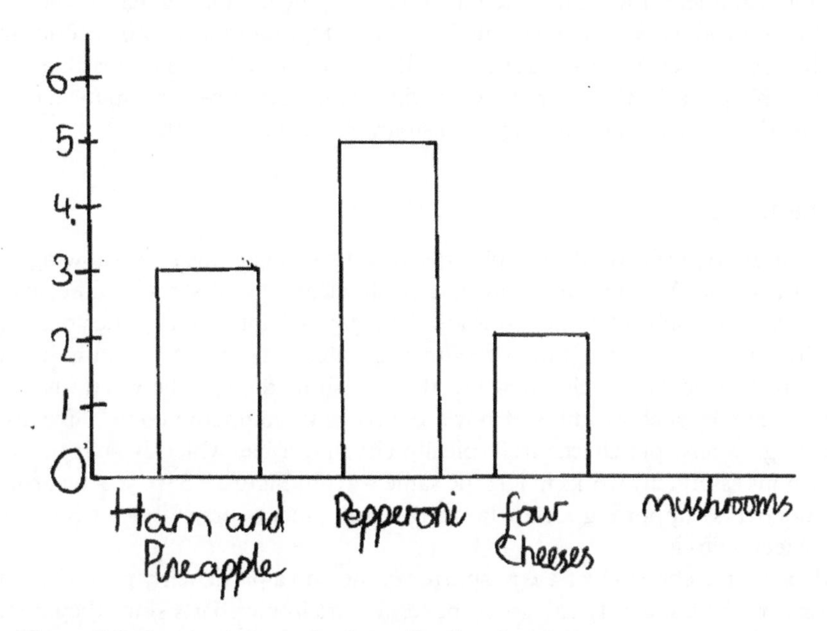

Figure 8.2 Bar chart displaying favourite pizza toppings

You will note that both the charts representing favourite pizza toppings have been drawn so that the blocks and bars are not touching. This is a display convention designed to show that it is categorical data. Whilst many sources suggest that all such data should be displayed in this way (separate bars), other sources suggest that categorical data with a clear order to it, such as numbers of children having school dinners on different days of the week, can be presented with the bars touching. An area of contention!

The labelling of the y axis on a bar graph in Figure 8.2 was an example of a continuous scale and this means that types of bar graph can also be used to display continuous data. For practical reasons this is often grouped. For example, if we wanted to graph children's heights we might opt to display numbers of children in given height bands with bars drawn touching to indicate the continuous nature of the data. From a practical perspective each child's height would probably be recorded to the nearest centimetre, which essentially turns continuous data into data which can be dealt with in the same way as discrete data, but the construction of the graph will show clearly that the data is measurement-orientated. Note that height bands should be very carefully constructed to avoid overlap between them; for example, $120 \leq x < 130$ and $130 \leq x < 140$; otherwise it would be problematic if one band ran from 120 cm up to and including 130 cm and the next from 130 cm upwards if a child's height was recorded as 130 cm!

A graph which children will probably come across later in their education is something referred to as a 'histogram', a particular type of bar chart which allows us to deal appropriately with bands of different sizes where these are necessary as part of the research. Here the area of the bars is varied to reflect differences between the categories. For example, if I was interested in comparing the hobbies undertaken by school-age children and working adults I might use a histogram to take into account the fact that the age brackets will vary quite a bit. If I ignore this, it is quite possible I will not get a true picture as to the extent to which different age groups participate in particular hobbies. For more on the workings of histograms see Cooke (2001).

Pie charts

Pie charts are typically used to display discrete data and are ideal for showing proportions of a whole. Visually they can be a really clear way of showing information as long as there are only a few categories to the data; with lots of categories they become unwieldy and their interpretation requires repeated reference to a complicated key. They work particularly well where we want to show the split between just two different categories such as ratios of boys to girls or to compare people's like/dislike of something. Whilst pie charts are typically circular, a bar which is split up to show proportions essentially works in just the same way. In Figure 8.3 we see the proportion of respondents supporting a new housing development versus those who registered their objection to it.

Because pie charts illustrate proportions, they do a particularly poor job of representing zero values; empty categories never feature in pie charts since they do not get a slice of pie!

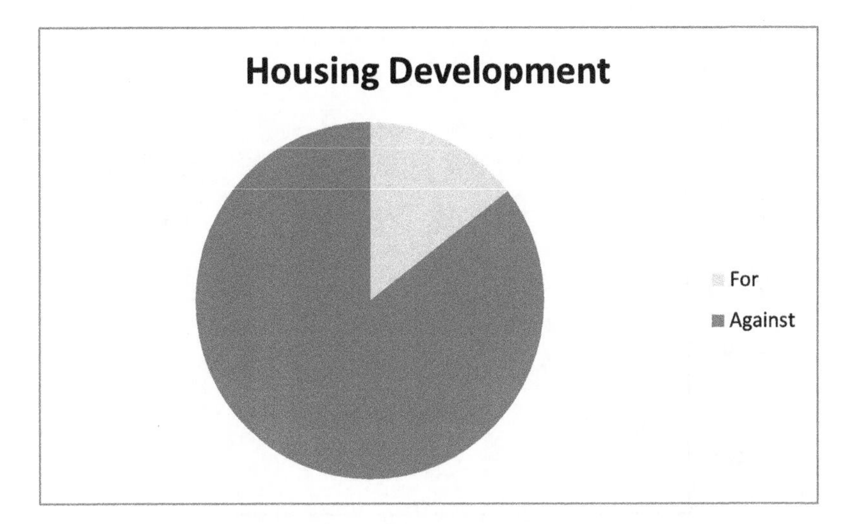

Figure 8.3 Pie chart showing local residents' views on a new housing development

Pie charts are traditionally associated with older children since drawing them by hand requires knowledge of degrees, ability to use a protractor and some quite advanced arithmetic skills, but the advent of computer software made making pie charts much more accessible and they are relatively straightforward to interpret. For more on their construction by hand, see Self-Test Question 4.

Scatter diagrams

Various terms for this type of graph seem to be used interchangeably: 'scatter diagram', 'scatter graph' and 'scatter plot'. In many ways the word 'plot' is quite useful since this type of graph is best used to illustrate the relationship or correlation between two ordered sets of data by plotting points on a single graph. Note that the purpose is comparison, therefore there will always be two sets of data and, for the graph to make any sense, it has to be ordered data.

For a scatter graph to suggest that two variables seem to be related, the points plotted appear to cluster around a line. For example, we would expect a positive correlation between children's age and their height; as children grow older they grow taller! The points plotted in Figure 8.4 show this clearly, with most children clustered around a line with just a few children who are particularly tall or short for their age. This is an example of continuous data being compared, but discrete data can also be compared, such as one set of test scores plotted against another.

Examples of negative correlation are less common, but plotting something like the temperature of a range of hot drinks as they cool over time would result in a scatter graph where the points would cluster around a diagonal running in the opposite direction to that in Figure 8.4.

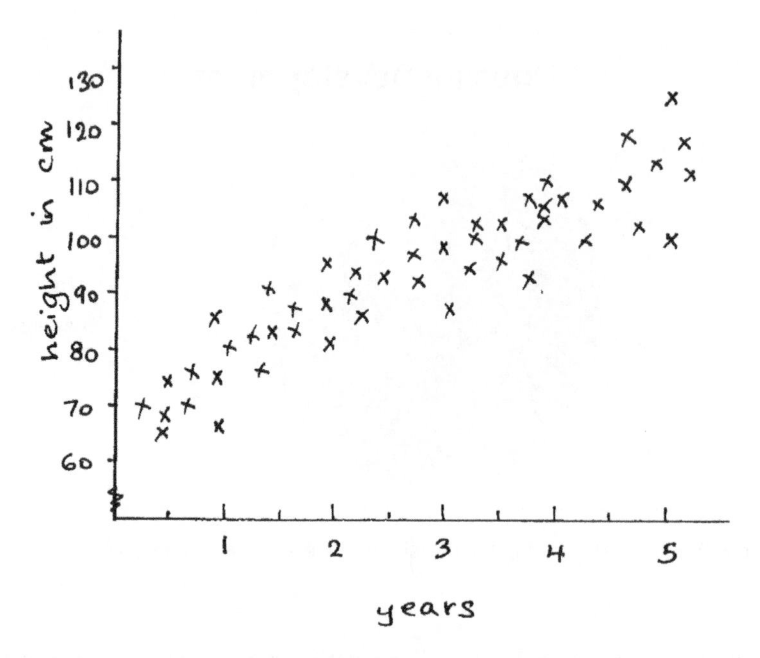

Figure 8.4 Scatter diagram showing the height of girls aged 0 to 5 years

There are some links between scatter graphs and conversion charts; for example, if we plotted one currency against another, the chart could be used to get a rough idea of how much something would cost us if we went abroad. Figure 8.5, based on the rate of exchange for the pound against the US dollar at the time of writing, shows a continuous line rather than a scattering of points because currency operates on a ratio scale. As continuous data there are values all along the line and the line could be extended beyond the data we see here; thus any amount of money can be converted using this approach.

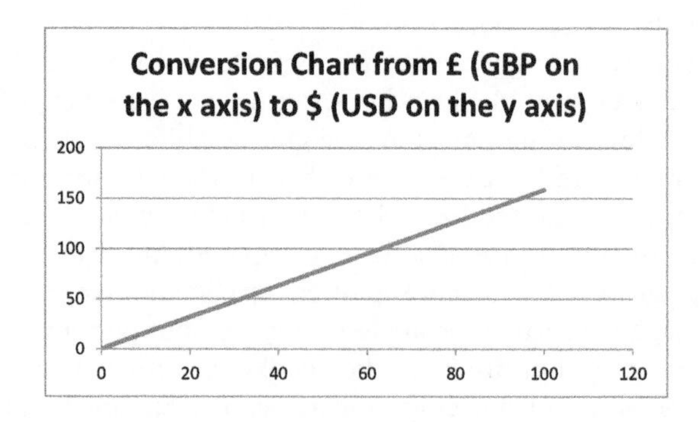

Figure 8.5 Currency conversion graph

Line graphs

Perhaps the best-known type of graph for displaying continuous data is a 'line graph'; this enables us to show how something changes (or stays the same) over time, with time conventionally labelled along the horizontal axis.

Taking temperature measurements, say, every hour on the hour, provides ideal line graph data. Measurements are plotted as points on a graph with the time running steadily along the x axis and the temperature shown on the y axis. The points can be joined to show the continuous nature of the data; we do not have measurements for the times in between the hourly intervals but we know that there must have been a temperature at all points in between and have to work from the assumption that it will not have fluctuated greatly. Rather than the slightly curved lines we see in Figure 8.6, the points would ideally be joined by straight lines.

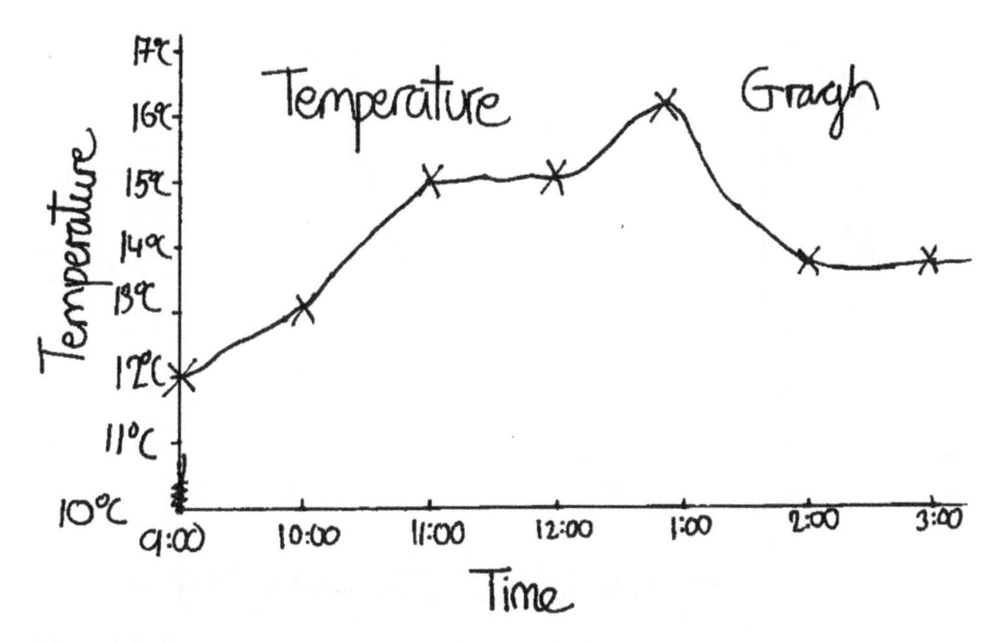

Figure 8.6 Classroom temperature over the course of the school day

If the temperature has risen from 13°C to 15°C, as we see between 10:00 and 11:00, this increase is reflected in the line rising from left to right. The decrease in temperature after 1:00 is reflected in the line dropping sharply away; perhaps the weather turned cloudy! Whilst gradients of lines help to suggest how steeply the temperature is rising or falling, it should be noted that comparisons between temperature charts can be made more difficult if axes are spaced differently. The rise or fall in temperature can be made to look more dramatic in one purely because of the design of its axes, with information presented misleadingly.

Problems also occur where data is missing and this is not reflected in the labelling of the axes. Imagine a class growing runner bean plants and measuring them each day to chart their growth. In Figure 8.7 the creator of the graph has left Saturday and Sunday off the axis because nobody was there to measure the height of the runner bean plants over the weekend, but as a result it looks like the plant has had a sudden growth spurt! Had all the days of the week been shown on the graph we would actually have seen that this particular bean grew very steadily!

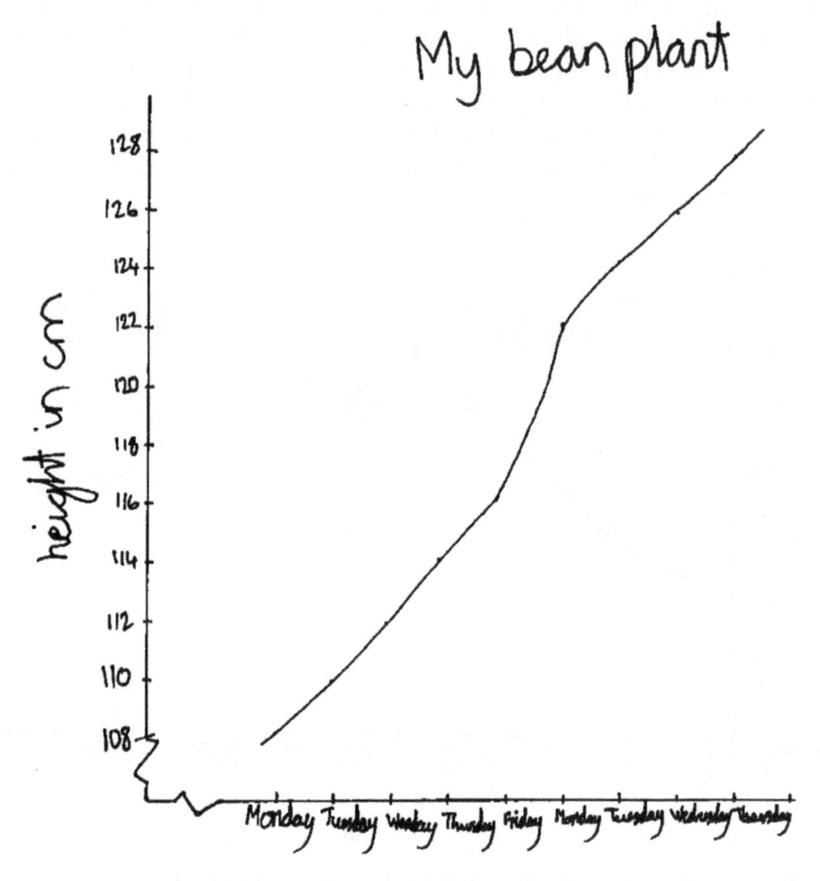

Figure 8.7 Inaccurate line graph showing the growth of a bean plant

Figure 8.8 is an example of an occasion when a line graph would not be the most appropriate choice since information about the number of children cycling to school during the summer term is discrete data and would be better displayed as a bar chart or similar. The values plotted were the right ones: two children cycling during April, six in May and so on, but joining those points is nonsense as the values on the lines between the points have no meaning. The number of cyclists does not decrease gradually from July to August; the data just tells us that there were five children riding to school in July, and none in August as it was the summer holidays. The only possible exception to this is a time series graph designed to show trends over a longer period of time.

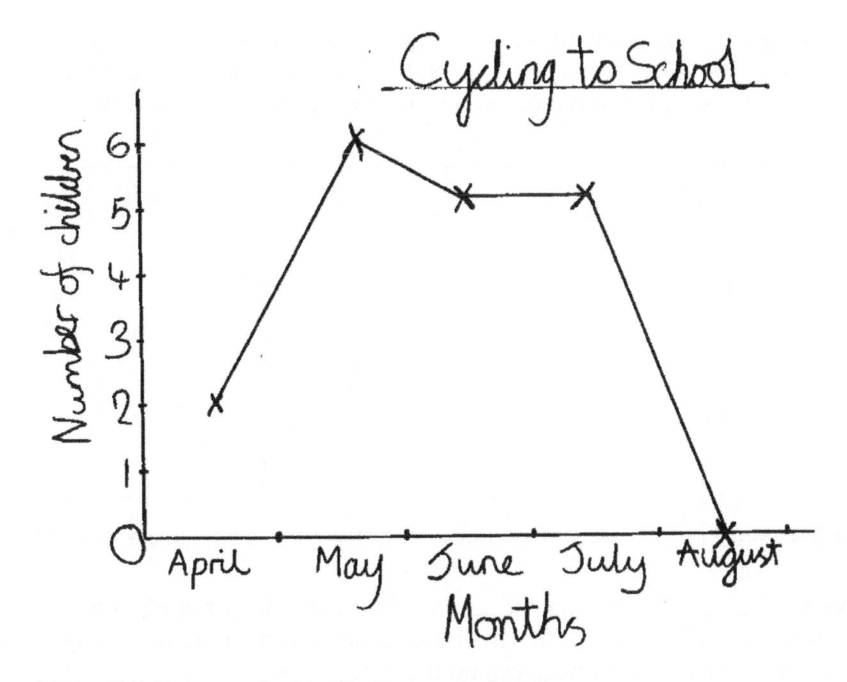

Figure 8.8 Data unsuited to display in a line graph

Common misconception

'The line graph shows that the car drove up and down hills'

Misinterpretation of line graphs is common in the early stages; for example, rather than interpreting the changes in speed as the car travels (acceleration and deceleration) interpreting the slopes or gradients as hills.

Sorting diagrams

Just as lots of real-life topics and other curriculum areas provide rich sources of data to be sorted, so do topics within mathematics, such as being able to sort shapes or numbers. There are various different sorting diagrams that can be used both for the sorting process and to display the findings.

Carroll diagrams, an example of which we see in Figure 8.9, sort by either one variable or two. Our variable in this instance is whether a number is prime or not; thus the single variable gives rise to a two-part diagram. Four-part diagrams (see activities 8.10 and 8.11) arise where two variables are considered together. Either way, the diagram operates such that all possible data should be able to be encompassed within the diagram and this is achieved by ensuring that whatever criterion is chosen for one category, the other side of the line is labelled as the opposite. So if one category says 'red' the other side must say 'not red'; blue is not a suitable alternative otherwise something yellow will have nowhere to go! So in Figure 8.9 'prime' is partnered by 'not prime' rather than (say) 'odd', which would give us a problem for all even numbers other than 2.

prime	not prime
7 2	81
13	27
5	56
11 23	32
97	74
	50

Figure 8.9 Carroll diagram for sorting numbers which are prime from those which are not

Venn diagrams operate slightly differently; the simplest example is a single circle or set as we see in Figure 8.10. If something does not belong in this, or any other set, then it can be placed somewhere around the outside.

Figure 8.10 Simple Venn diagram for sorting farmyard animals

Overlapping sets are possible with Venn diagrams; two sets intersecting where the categories potentially overlap. See, for example, activities 8.10 and 8.13.

Some computer software provides opportunities for children in the early primary years to practise their sorting skills using such diagrams. Users need to be conscious, however, that the quality of the mathematics is dependent on the mathematical knowledge of the person who created the materials. I have come across software which reinforces misconceptions, for example, by forcing squares and rectangles into entirely different categories, when in fact a square is a special type of rectangle; for more on this, see Chapter 10.

Tree diagrams provide another sorting mechanism in which questions are asked which require only 'yes' and 'no' answers. It is about learning to ask good questions, for example, that would be able to identify particular shapes or types of plant. In Figure 8.11

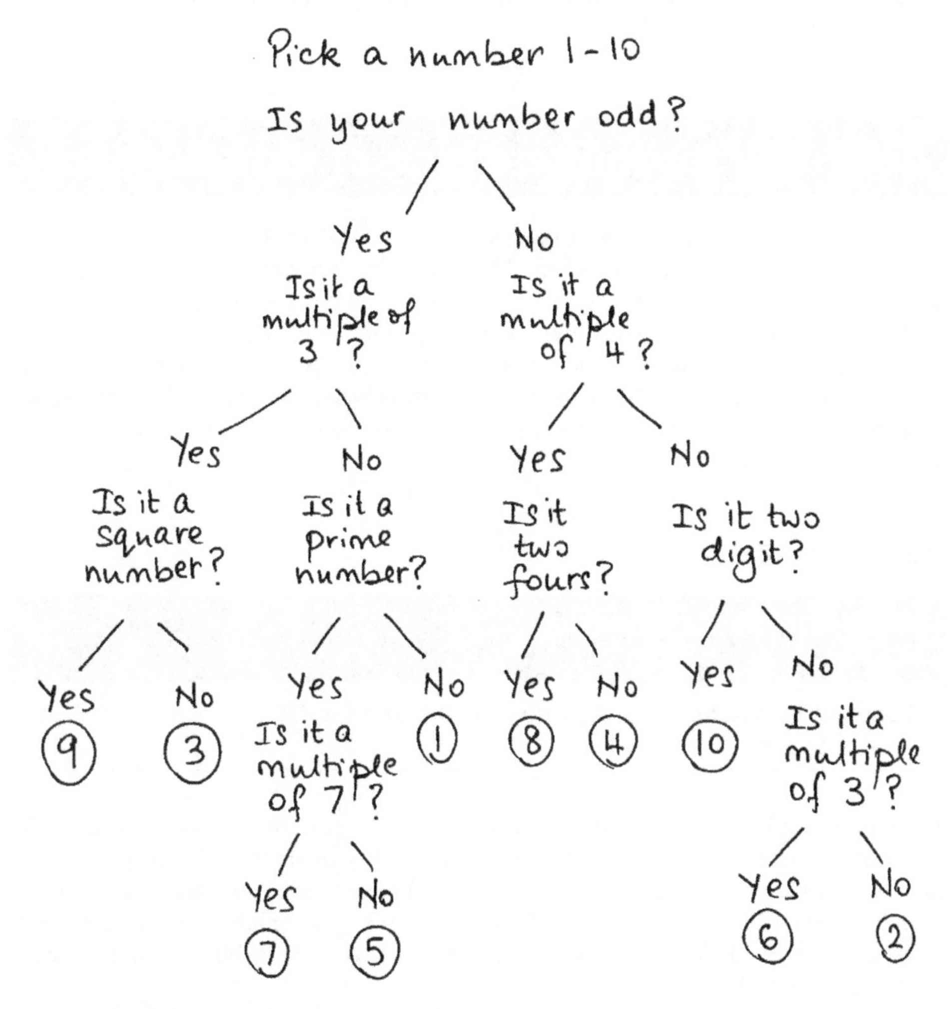

Figure 8.11 Tree sort for numbers 1 to 10

we see that someone has attempted to identify all the numbers from 1 to 10 using this approach, testing to check that it would work reliably. What other questions could have been asked? Could the order of the questions be altered? For additional ideas, see activity 8.12.

Note that early sorting occurs naturally in the primary classroom, particularly whilst putting things away: all the toy cars belong in the garage; the Noah's ark animals do not belong in the farmyard, and so on. Such equipment can, however, be incorporated purposefully into data handling lessons, even with older children. Reminiscing with the children, 'do you remember when you used to play with...?' can be used to initiate data handling activities.

Clearly there are other types of graphs and charts you will meet during your teaching career, such as those for displaying educational data relating to cohorts of children. For further information on such topics, see Haylock (2010) or Mooney *et al.* (2012).

Activities for the teaching of data handling

Some of the following activities target generic data handling skills whilst others focus on particular types of graph or chart. Data handling activities often work well with children of different ages, but schools will want to be clear about the progression of skills they wish to target with different age groups and abilities to ensure that the children's learning is moved on. Where appropriate for a new type of graph, you might consider starting by asking the pupils to arrange themselves in a particular way to illustrate how the graph will look, or to physically move objects around to make counting and comparison easier.

Pupil activity 8.1

Self-portrait for data handling
Y1 to Y3

Learning objective: To appreciate that the same data can be displayed in different ways.

Asking the children to draw pictures of themselves on squares of cardboard provides the starting point for this data handling activity. My example is based around the months in which the children are born but could equally focus on personal attributes such as hairstyle or eye colour, or where pupils live. The children, with help where necessary, attach their pictures to the whiteboard to make a block graph showing everybody's birthday month:

6									☺			
5					☺				☺			
4			☺			☺				☺		
3	☺		☺		☺	☺				☺		
2	☺	☺	☺		☺	☺			☺	☺		
1	☺	☺	☺		☺	☺	☺	☺	☺	☺		☺
	Jan	Feb	Mar	Apr	May	Jun	Jul	Aug	Sep	Oct	Nov	Dec

The graph is labelled together and discussed, noting:

- the month or months in which the most children were born;
- the month or months in which the fewest children were born;
- any months in which nobody was born;
- any months with the same number of children as each other;
- differences between numbers of children in various months;
- numbers of children represented by the graph in total.

Highlight a few individual pieces of data within the overall data set by commenting on the child who has a birthday around Christmastime, say, the idea being to fix some of the information in the children's minds.

The graph then needs to be left where it is until you have the opportunity to rearrange the data without the children present, for example by displaying the months down the left-hand side of the board. I have always found that the element of surprise appeals to the children; they come back in and the graph has changed! But of course the key learning is that exactly the same data is still presented, and this will need to be made explicit in discussion. If we went through the bullet points above, we would get the same answers.

Whilst the children are only being asked to provide one piece of personal data, which could be considered quite a low level of demand, the emphasis is on recognising that the full data set can be displayed in different ways without altering any of the individual facts. Note the cards should be squares of the same size to aid clear display of the data.

Pupil activity 8.2

Observations recorded
Y1 to Y4

Learning objective: Ability to make a reasoned choice as to the best way of presenting observed data.

Whilst the example here is a pictogram based on children's observations of the different types of birds visiting the school garden over a period of 20 minutes, the children could explore different ways of presenting this data by trying out different graphs on the computer. Key questions are:

- Which types of graph work and which do not? Why not?
- Of the types of graph, which are OK? Which might be considered the best choice and why?

Discussion will support children's formation of ideas and they could be asked to vote for their favourite style of graph, justifying their choice. Other observational data would work equally well of course.

Pupil activity 8.3

Purposeful pictograms
Y2 to Y5

Learning objective: To understand how to present large amounts of data in a pictogram.

Because pictograms sometimes use pictures to represent more than one of something, children need to deal with large enough quantities of data to warrant this. An example I have used successfully in the past is to ask children to count how many books they have in their bedrooms, knowing that this will generate a range of data. Because large numbers are involved, this gives a good rationale for letting a little picture of a book represent 10 books. Multiples of 10 work particularly well because this models the structure of the number system and the children can be asked to suggest solutions for numbers of books which are not exact multiples of 10, such as rounding them up or down or drawing partial pictures.

This could be extended to comparisons between numbers of books in different local libraries, or perhaps sales data comparing numbers of real books sold versus electronic copies.

Pupil activity 8.4

Don't get caught out!
Y1 to Y6

Learning objective: To develop ability to interpret different types of graph.

One way of enhancing children's ability to interpret graphs and charts is to present examples with three statements made about each one: two true statements and one false statement per graph. Children work collaboratively to agree which statement they think is false, as we see in the example below. Asking the children to justify their choice of false statement will help you to access their thinking and assess their ability to interpret the different types of graph. Here the first statement is false as the graph suggests there is a relationship (albeit a weak positive correlation) between size of hand span and length of middle finger. Working collaboratively and talking with each other will help the children to feel well prepared to argue their case.

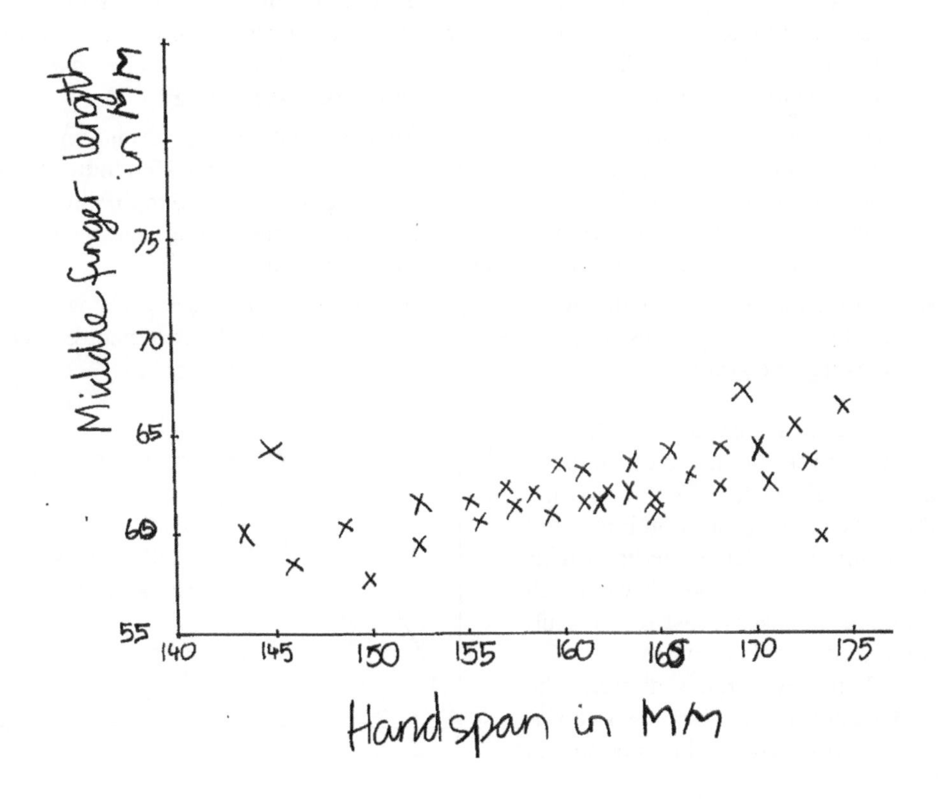

- Hand span and middle finger length are not related in any way.
- Both measurements are given in millimetres.
- There's somebody with quite a long middle finger (about 64 mm) but quite a narrow hand span (about 145 mm) in comparison with the other children.

Pupil activity 8.5

Follow the line
Y5 to Y6

Learning objective: Ability to interpret a line graph.

Ability to interpret line graphs seems to come with practice, quite probably making mistakes along the way. Once children have been introduced to the idea that line graphs chart change over time, a good way of providing experience is to tell stories which involve this and ask the children to visualise the line going up and down (or remaining level) as you tell the story. Differentiation is possible through keeping the stories simple or making them more complex. For example:

- Five days before going on holiday I decided to try to lose weight. I ate less and for the first three days my weight (mass!) stayed the same but over the next two days I lost 1 kg. On holiday I enjoyed lots of lovely food and my weight crept steadily up. By the end of the week I was 2 kg heavier than I was at the start. (The unknown starting amount potentially makes this scenario more challenging.)
- The speed of my journey home from the airport was interesting. It was stop start on the motorway and I never got above 40 miles per hour. After 20 minutes I left the motorway and once I was on the minor roads kept a steady 60 mph other than going through the villages with 30 mph speed limits. (Quite ambiguous in parts to encourage children to interpret the 'line' in their own way.)
- Once I was home I decided to record the temperatures for a week.... What story could you tell about this scenario? What other scenarios would lend themselves to creating line graphs?

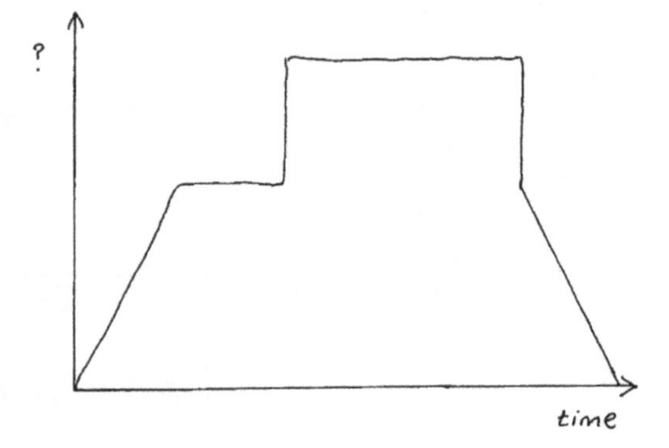

After each story give the children a minute to sketch how they think the graph would look before comparing their sketch with a neighbour and discussing the relative merits of their pictures. Mini whiteboards work well for this as it makes it easy for the children to change their minds.

Then give the children the opportunity to tell their own stories about various lines. This is mine and was drawn to represent the 'story' of a bath filling with water. You can probably tell the points at which the person got in and out!

Pupil activity 8.6

Focus on scales
Y3 to Y6

Learning objective: Ability to read scales accurately in order to interpret graphs.

Environmental topics often display data in different types of graph, such as information about levels of recycling. But actually it is not the topic that is important, but finding a range of graphs where the children need to interpret different number scales in order to appreciate the data (you might try data about music downloads in contrast). Having discussed the information available in the different graphs:

- ask individuals to make three statements about what they have found out (the accuracy of the statements will help you to assess the children's ability to interpret scales and data);
- ask groups to suggest which of the graphs, if any, they would present differently to communicate its information more clearly.

A more advanced focus on scales is to explore the different impressions given when the same data is presented in graphs with different scales.

Pupil activity 8.7

Easy as pie!
Y1 to Y6

Learning objective: To understand data presented as a pie chart.

Because computer packages allow children to generate pie charts so easily, this exposes children to them from quite a young age and they can begin to interpret simple data presented in this way. Even getting into a circle standing with people who belong in the same group as you (because they are also wearing sandals or whatever) gives an approximation of a pie chart (we see that lots of people are wearing trainers and hardly anybody is wearing sandals like us).

Pie charts are best used to illustrate just a small number of categories, and simple survey data where answers are only 'yes', 'no' or 'don't know' is ideal. Whereas older children might gather each other's views on contentious topics ('Do you think animals should be kept in zoos?'), with younger children an easier option is to ask a 'Do you like...?' question. For example, if we ask 'Do you like prawns?' we are likely to get a range of answers and these can be displayed and discussed in a three-part pie chart. Another option is to work with scenarios that lend themselves to 'yes, always', 'no, never' and 'sometimes' answers to questions such as 'Do you like swimming?' and children can begin to discuss whether they think they could generalise from their limited sample to draw conclusions about the wider population.

Following this, a nice extension is to present children with a pie chart with the title and key missing; given the sizes of the 'slices', what do the children think it might represent? What might the actual numbers of people be? Beginning to appreciate the size of each part in relation to the others is a valuable precursor to later work constructing pie charts by hand and needing to understand that each part shares a portion of the full 360°.

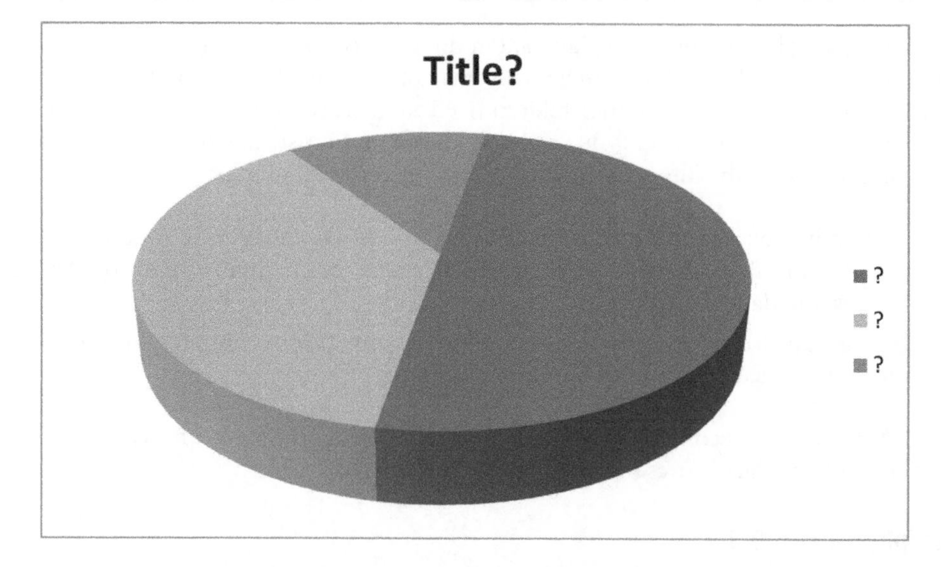

In Chapter 2 we met the idea that data gathered from surveys can be really useful for modelling an application of percentages; this can be linked to the generation and interpretation of pie charts. Remember that 100% is the maximum percentage possible in this case as we are not talking about percentage increase or decrease.

Pupil activity 8.8

Vehicle data
Y2 to Y6

Learning objective: Ability to plan carefully in advance how to collect appropriate data.

The starting point for this activity need not be to find out about traffic data, but does need to be a scenario where the data could potentially need to be gathered at speed. Choice of topic clearly needs to be made with health and safety in mind; some schools are in locations where gathering traffic data would not be feasible. But, assuming it is possible, set the scene by discussing the possibility of a new crossing patrol or retaining an existing operation. To prove the need for the patrol, the head teacher would like some data about the amount and type of traffic passing the school. Once the children have had a chance to discuss this, ask them to shut their eyes and imagine standing with their clipboard as the traffic passes; this may prompt them to consider ways in which the recording of the data can be made as slick as possible so that they can keep up with it. Depending on the age of the children, you may want to guide them towards certain techniques such as tallying or ticking predetermined categories. Discussion may also result in some suggestions, such as different children recording different types of vehicle, or only focusing on traffic on one side of the road with another team dealing with the other side. Once children have designed their data collection sheets, these should be tried out, and this may result in changes being made before using them again to collect a final set of data. Finally, the appropriateness of the data can be tested by showing it to the head teacher (or another willing adult) to see if the children have provided the requested data on both numbers of vehicle and vehicle type. Graphical presentation could be incorporated as part of the activity, and for the older children this type of activity could form part of a more extensive investigation into traffic flow, car parking, and so on.

Pupil activity 8.9

Keeping an eye on the weather
Y4 to Y6

Learning objective: Ability to use graphs and charts confidently to compare information and inform decisions.

Data handling relating to the weather can be about either gathering primary data (such as millimetres of rainfall overnight or playground temperatures) or accessing secondary data (such as weather information for various countries from websites such as http://www.weather-and-climate.com/). As an avid fan of travelling, such weather charts help me to decide not only the best times of year to travel, but also what sorts of clothes I might need to pack. Rather than giving pupils free rein to investigate, it is probably best to channel their investigation by letting groups each choose three destinations to research and compare. I would probably group some different destinations for their potential as safari, lazing on the beach or historical sightseeing holidays, making sure that the choices are quite diverse. Having a particular type of holiday and time of year in mind will help to give some focus to the exploration of the data. At the end of the process they should be able to complete the sentence:

I'd choose to travel to................ because.................

It is the ability to justify their answers that is important, and children could also be asked to suggest what their second choice would have been and why. The research may also give rise to other questions and considerations and a little background knowledge can help this along, so you may benefit from a little research of your own. For example, it might be that rainfall seems high, but is typically a short, sharp shower at a predictable time of day, and not enough to put you off a beach holiday!

Pupil activity 8.10

Sorting out numbers
Y1 to Y6

Learning objective: To appreciate mechanisms for including and excluding data in sorting diagrams.

One of the benefits of working with both Venn and Carroll diagrams in the same lessons is that the children will be able to develop their understanding of how each of them work through realising that there are similarities and differences. Clearly being asked to articulate what they notice will play a key role in enhancing their understanding. Those we see here have been used to sort numbers which are multiples of 3 and/or 5; whilst learning about the types of diagram, hopefully the children are also reinforcing their learning about multiples! The choice of number topic is likely to vary in accordance with the age and ability of the children; with very young children you might want to try just a Venn diagram with ambiguous sets for 'big numbers' and 'small numbers' which may well prompt the overlap of the two sets for numbers causing the most discussion!

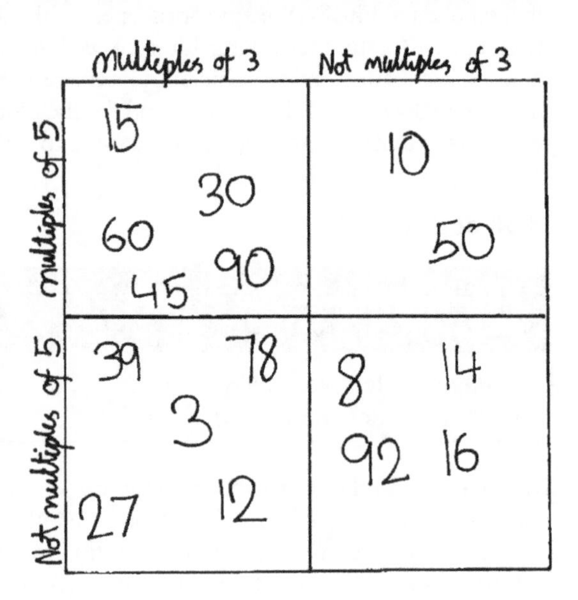

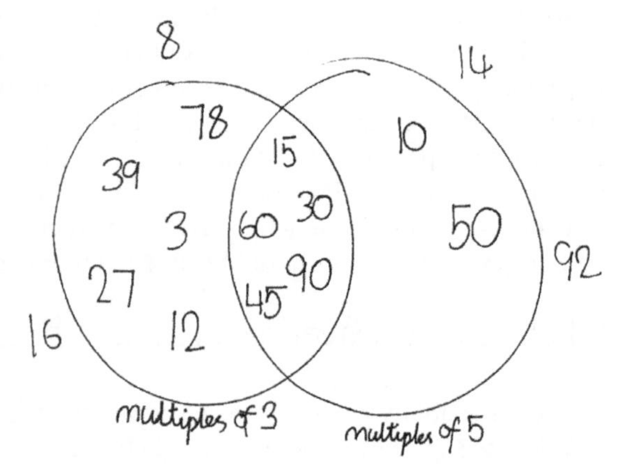

Having completed such diagrams, children can be encouraged to make more general statements about the data. Sorting multiples of 3 and 6, for example, should prompt a general statement about one group being inclusive of the other.

Science offers many alternative topics (some of which are explored in activities 8.11 and 8.12) such as sorting plants, mammals, rocks or foodstuffs.

Pupil activity 8.11

Carroll diagrams for categorising Y2 to Y4

Learning objective: To appreciate that Carroll diagrams provide a useful mechanism for categorising things according to two criteria.

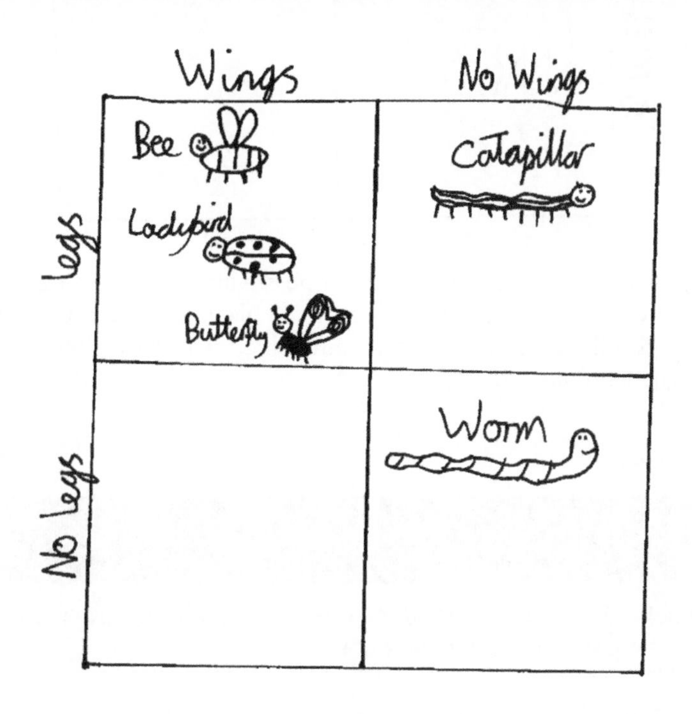

I have always found Carroll diagrams a really useful way of sorting different sorts of creatures in science, in particular to distinguish between insects and other minibeasts as we see in the diagram here. The empty region tells us quite a lot about creatures: we do not get any with wings but no legs; just imagine the issues when coming in to land!

Another topic which works well is to gather and record data about people's drink preferences. Here we see how various members of staff take their coffee! This tells us exactly how many to make of each type and suggests we will not need any black, no sugar.

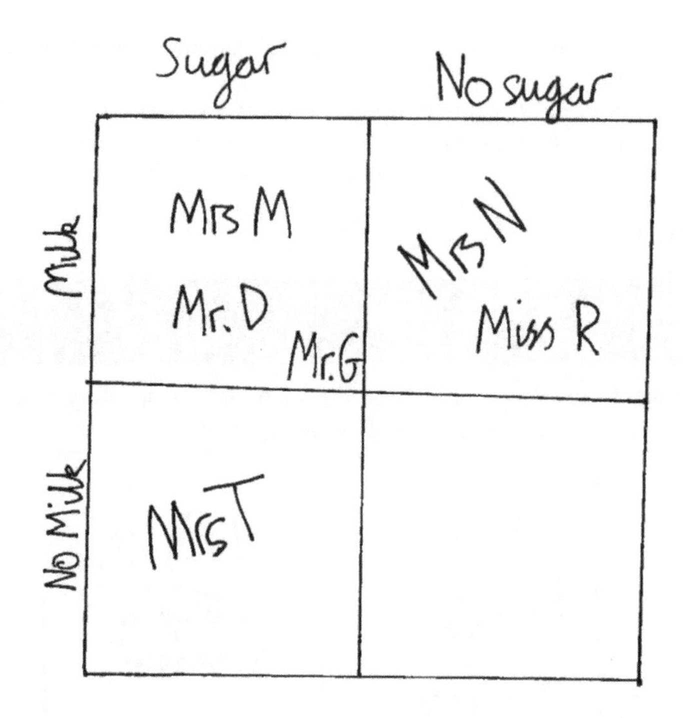

Working with just one criterion is also a possibility, such as using a Carroll diagram to record whether something floats or not. This might give rise to adding a second criterion if a child notices or suggests something, such as whether or not items are made of a particular material.

Pupil activity 8.12

Yes or no?
Y2 to Y6

Learning objective: To understand the role of questions in sorting and categorising data – in particular, scientific data.

Earlier in the chapter we saw a basic sorting mechanism, a 'tree' for identifying different numbers from 1 to 10. Here we consider the categorisation of living things using a similar approach. Even quite young children can use such diagrams with guidance since it can focus on the answering of simple questions such as:

- 'Does it have a flower?'
- 'Is the flower yellow?'
- 'Does it have four legs?'
- 'Does it have fur?'

But categorising things, particularly data from the natural world, is a complex task even for scientists. So rather than expecting children to tackle such a task in full, the number of plants or animals needs to be limited to a few examples (such as spring flowers or farmyard animals) and structured in such a way that it becomes manageable. One way of doing this is to ask a carefully chosen starting question and for the children to concentrate on just one of the 'branches' leading from either a 'yes' or 'no' answer. A couple of good starting points are:

- 'Has it got a backbone?' with 'yes' leading towards categorising mammals, fish, birds, reptiles and amphibians.
- 'Does it lose its leaves in winter?' with 'yes' leading to an exploration of the deciduous trees in the school grounds.

Published field guides on the classification of plants and animals may prove useful in supporting what is partly a mathematics lesson but also very much a science lesson.

Pupil activity 8.13

Venn diagrams with a cross-curricular focus
Y1 to Y3

Learning objective: To understand the contribution of Venn diagrams to knowledge in other curriculum areas.

The two examples given here are drawn from science (materials) and history (toys), but you will no doubt have other ideas you could add. The emphasis is on using a Venn diagram or diagrams to investigate and better understand the topic.

A focus on materials can be initiated through exploring kitchen utensils since these are typically made from a variety of materials and include examples which incorporate more than one material. We would omit sharp knives for obvious reasons! I would start the lesson by asking the children to carry their item around the classroom, talking to each other about similarities and differences between the items. Encouraging descriptive language is important since careful observation and the ability to distinguish between features are the basis for sorting. The utensils would then be sorted into hoops according to what they are made of, with items made of more than one material prompting the need for overlapping sets. Ideally you would also want at least one set to remain empty, such as paper. The final stage is to consider what the sorted information tells us and to draw some conclusions about why items might be made of certain materials and not others, or why somebody might choose a wooden, plastic or metal spatula.

With a history topic, a Venn diagram allows us to note similarities and differences with the past, and toys prove a suitable topic, since some things (dolls, balls, puzzles) have been played with children across the ages whereas others (electronic games) are a modern invention. The act of sorting these prompts us to think about why this is the case (which, like the activity above, might be to do with materials) but could also engender empathy with children of the past.

Pupil activity 8.14

Checking the Census
Y5 to Y6

Learning objective: Ability to interpret discrete or categorical data, organising and displaying it for the purpose of communicating findings clearly.

Accessing historical data provides some nice opportunities for children to answer questions about the past through following the data handling cycle, such as exploring the trades and other jobs that people were registered as doing. Sometimes picking topics with lots of variety to the data is helpful as it forces a need to think about grouped data in order to present it in a manageable way.

Bear in mind that any generalising ought to be done cautiously; the spread of trades in one population sample will quite possibly vary from those of people in another locality.

Pupil activity 8.15

Olympians
Y4 to Y6

Learning objective: Ability to interpret continuous data, organising and displaying it in order to communicate findings clearly.

The Olympics and other sporting events naturally generate a wealth of (secondary) data, and the emphasis for this activity is on exploring information about 'how far' and 'how fast'. Note that both of these are examples of measurement data and are therefore continuous. Given a table of data, the children should work in pairs to decide how this information could be presented graphically, bearing in mind that the continuous nature of the data needs to be taken into account and that the graph needs to present the data accurately. This stage might involve trying various graphs produced from a spreadsheet of the data. The children's final choices should be prepared as part of an information sheet, and this might include making statistical statements such as referring to averages. Pairs should be prepared to explain their information sheet and answer any questions.

Primary data could also be used from PE lessons with children choosing between entering the 100 m with their time measured to the nearest second using a stopwatch, or attempting to throw a quoit the furthest, with the distance measured to the nearest centimetre. This has the advantage that the children can be involved in collecting the data first hand and having to deal with it in its raw state before going through the same processes described above.

Pupil activity 8.16

Types of average
Y5 to Y6

Learning objective: To understand the way in which different types of average are calculated.

A good way of exploring mean, median and mode is to either generate or provide some simple results from games played by an individual. The set of data needs to be numerical, so ideally a game or activity which gets a score such as rolling two dice and adding, achieving the following scores over ten successive games:

8, 8, 7, 7, 3, 8, 4, 9, 11, 7

In Chapter 9 we will think about whether we might have anticipated a range of scores such as this, but for now we are interested in helping children to get to grips with how to work out the mean, median and mode:

- To work out the mean, we need to add all the scores (the sum is 72) and then divide by 10 (the number of scores we have), giving us a mean of 7·2.
- For the median it would help to see all the scores in order. We find that the values in the middle of the list are 7 and 8, so, splitting the difference, we get a median of 7·5.

  3, 4, 7, 7, 7, 8, 8, 8, 9, 11

- The mode is the score occurring the most. We find that there are two modes in this case, 7 and 8, which both appear three times.

Rather than children calculating the mean, median and mode, a simple but really nice alternative is to provide each of the averages (suppose they are all 6) and ask the children to generate possible data sets to match these.

Whilst both activities are concerned primarily with the calculation of mean, median and mode, it is essential that children also begin to understand the relevance or irrelevance of the different types of average (see activity 8.17). Discussing how well the three types of average represent the data (quite well in the case of my dice scores) is a good start!

Pupil activity 8.17

Purposeful use of averages
Y5 to Y6

Learning objective: To understand the importance of choosing types of average with care.

This activity assumes that children already know how to calculate different types of average, focusing their attention instead on when it might be appropriate to use one type rather than the others to represent the data in a sensible way. As well as the spread of actual data, the choice of average should also be affected by the nature of the data and why someone wants a measure of central tendency. Exploring lots of different scenarios together is the best way of achieving this, for example:

- A shoe shop owner who has data on shoes already sold and who wants to know whether averages can help with restocking. [Given the nature of shoe size data only the mode would be a suitable average. This could prove useful; if data suggests the shop has sold more size 39 shoes than any other size, then having more of those in stock might be a good plan.]
- A company owner would like to be able to talk about average wages paid. [With employees at different points on the pay scale, any modes could be coincidental, so mean or median are likely to be better choices of average. If there are any extreme results, as we had with the pocket money earlier in the chapter, then the median may be a better reflection of central tendency.]
- An electricity provider wants to be able to tell customers how many units of electricity are used on average per day. [Whilst the results will vary from household to household, with lots of results and no particular extremes, the mean would probably be a suitable average.]

What other scenarios could you add to illustrate choice of average?

Displayed around the room would be the terms 'mean', 'median' and 'mode', and after hearing each story children might be asked to point to the one they feel (gut reaction) would be the most appropriate choice. They should then be given enough time to discuss their ideas with each other and to work out any of the averages they feel they want to check (assuming the stories are accompanied by data), before committing to a final answer and being prepared to justify it. Open discussion of choices will be key to recognising that averages chosen need to be suitable.

Pupil activity 8.18

Real data or a fairy tale?
Y1 to Y6

Learning objective: Ability to select an appropriate diagram for displaying given data and to interpret the data appropriately.

This activity starts with a well-known fairy tale and the children have to take the raw data they are given and decide how best to present the data in a graph or graphs. For example:

- *Goldilocks and the Three Bears*. The children are told the starting temperature of the porridge when it is first made, and the temperature of each bowl when Goldilocks sits down to try it, eating Baby Bear's porridge all up of course! [Graphs could relate to the temperatures, but children might also find a way of showing Baby Bear's porridge decreasing in quantity.]

- *Cinderella.* The story starts with Cinderella working unhappily as a maid, and she becomes even more unhappy when she finds out there is to be a ball and she cannot go. She is elated when her fairy godmother says she can, but then worried she has nothing to wear. Hopefully the children know the rest of the story; how could they graph Cinderella's happiness? [A line graph is probably the best choice as her feelings change over time.]
- *The Three Little Pigs.* When the pigs are old enough they leave home and need to build their own houses so acquire straw, sticks and bricks; given the mass of each, how would the children present this information in a graph? [Could the children also consider designing a graph to give information about the wolf? Amount of puff required?]

This activity can be adapted for any age group by using different stories incorporating anything from simple, discrete data to complex, continuous data. Concept cartoons could also be created to stimulate discussion.

Pupil activity 8.19

School data
Y3 to Y6

Learning objective: To develop an appreciation of the full data handling cycle through purposeful inquiry.

On one occasion when I visited a local school the Year 5/6 children were exploring the data generated by the recent summer fête. Groups were investigating stalls of interest to them and thinking about issues such as: amount of preparation and number of people required to run the stall on the day; preparation and running costs; takings on the day (related to the number of visitors to the stall); cost of prizes if any; profit; ease of setting up and dismantling the stall; and the enjoyment of the people who attended the stall. These were complex sets of data with complicated interrelationships in play. Rather than shying away from the tough questions the children were expected to persevere with them and sure enough they did, eventually preparing a presentation (including tables, graphs and charts) and making their recommendations to the Board of Governors. Decisions made for the next fête will certainly be well informed!

Other school topics with similarly rich data can of course be used, such as an in-depth investigation as to the types of book the school library should stock, taking into account the views of children and teachers. Some topics may suit younger children better, such as investigating the most popular type of playground or sports equipment to purchase.

Summary of key learning points:

- children should experience the full data handling cycle where possible and use data to answer questions;
- sorting and categorising underpin much data handling;
- the same data can be displayed in different ways, but some forms of graph will suit certain data better than others, noting the different requirements of discrete and continuous data;
- data is very much a part of everyday life and links with real life and other curriculum areas should be made;
- when presented with data and statistics, these should be questioned rather than accepted at face value.

Self-test

Question 1

If we were to ask somebody how much television they watched yesterday, their answer (a) would be an example of discrete data, (b) would be an example of continuous data, (c) could be an example of either discrete or continuous data.

Question 2

Counting the frequency of each letter of the alphabet in the sentence 'count the frequency of each letter of the alphabet in this sentence' gives me the following table:

a	b	c	d	e	f	g	h	i	j	k	l	m	n	o	p	q	r	s	t	u	v	w	x	y	z
3	1	4	0	11	3	0	5	2	0	0	2	0	5	3	1	1	2	2	8	2	0	0	0	1	0

The mode is (a) the letter 'e', (b) 0, (c) 11

Question 3

Looking at the table in question 2, the median would be (a) n, (b) 1·5, (c) neither of these

Question 4

If I wanted to construct a pie chart (using a protractor) to show that 70% of people surveyed liked curry, 25% disliked it and 5% weren't sure, those who dislike curry would have (a) a 250° slice of the pie chart, (b) a 100° slice of the pie chart, (c) a 90° slice of the pie chart.

Self-test answers

Q1: (c) This question was worded fairly ambiguously on purpose. Strictly speaking, the fact that we asked a 'how much' and not a 'how many' question implies that the answer should be measurement-related and would therefore constitute continuous data. This would be an answer such as 'I watched television for about an hour and a half'. But what might happen is that we get an answer relating to the number of television programmes the person watched. This would be an example of discrete data.

Q2: (a) The categories were the letters of the alphabet and since 'e' occurred more times than any other letter, this category is the mode.

Q3: (c) As the categories in this example are not numerical, we are unable to calculate a median or a mean.

Q4: (c) Pie charts are useful to show proportions within a whole so would be ideal to show this data. Because the full circle of a pie chart consists of 360°, the 25% of the people surveyed (which is a quarter of the full 100%) would be represented by a quarter of the 360°. 360 ÷ 4 = 90 so 25% is equivalent to a 90° portion of the circle.

Misconceptions

'This tally tells us that 5 cars went past'

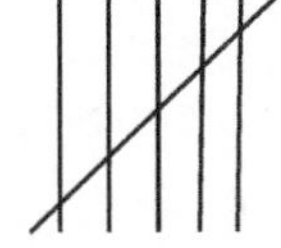

'There are 27 people in my class and I have asked everyone what their favourite fruit is: 10 children chose apples, 10 children chose strawberries and 10 children chose bananas'

'This graph shows clearly how many people like each crisp flavour'

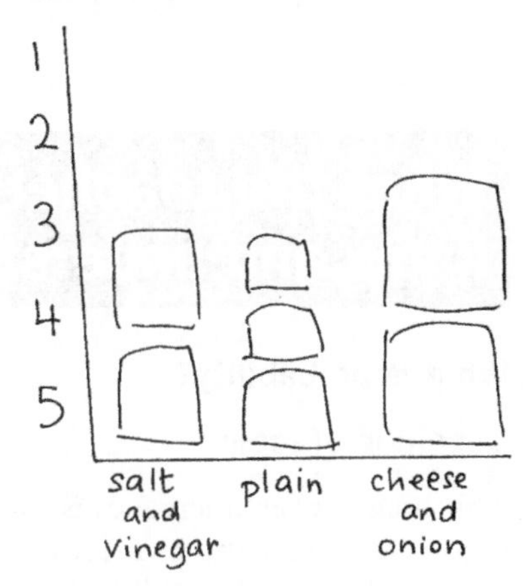

'As there are three cat pictures in this pictogram this must mean that three people have cats'

'The line graph shows that the car drove up and down hills'

9

Probability

About this chapter

This chapter builds on Chapter 8 by exploring the type of data generated in the context of probability and is organised into sections as follows:

- What is probability?
  - Definition of topic
  - Rationale for teaching probability
- Progression in probability
  - From the language of probability to the numerical conventions
  - Approaches to recording probability
  - Key terms describing aspects of probability

What the teacher and children need to know and understand

What is probability?

Definition of topic

This branch of mathematics is essentially about measurement of how likely an event is to occur, with the chance, or probability, being described using language or in numerical terms. Probability data exists in three main groups, all of which we ought to appreciate if we are to teach the topic effectively. These are introduced here and then explained in more detail below:

- statistical probability – probabilities generated by analysing previous sets of data, for example, based on information gathered from a small sample of the population;
- experimental probability – data generated through investigation;
- theoretical probability – anticipated outcomes derived mathematically.

In its most basic form, statistical probability is about learning from life's experiences and basing decisions loosely around what we think is most likely to happen. The woman who decides to put a spare pair of tights in her handbag when she's going to a wedding does so not because of any formal statistics on the topic, but because in her experience she has a habit of laddering her tights and thinks the probability of needing a replacement pair might be quite high – she wants to look her best! As well as likelihood influencing our daily decisions, statistical probability also affects most of us through insurance premiums since these are based on statistics regarding what is likely to happen. The underwriters clearly want to make sure each policy holder is charged enough; in other words, the company hopes to make a profit, receiving more money than it has to pay out! Costs are therefore calculated keeping relative frequencies in mind; if you live in an area prone to flooding, then your house insurance premiums are likely to reflect this. Remember that inaccuracies in statistics can occur when information is based on too small a sample, or where a sample is skewed in some way; therefore statistical probability suffers from some of the same issues.

Another example of statistical probability relates to weather forecasting. Whilst my decision about what to wear and whether to take an umbrella might be informed by wandering out into the garden and looking up at the sky, I doubt that weather forecasters take the same approach! We know they do not always get it right (how could they?), but they are making the best prediction they can based on the statistical information they have available to them, drawing on knowledge about typical weather patterns.

Well-chosen poems and picture books offer lots of opportunities for prediction and thinking about the likelihood of what will happen. In fact most stories, whether read, or watched in a film or TV drama (consider a good murder mystery) will have you thinking about what might happen next. This will be based on likely outcomes based on personal experience or what you have read or seen before; story lines with a twist work well because they surprise us and challenge our preconceived ideas! But whilst the outcomes in stories may come as a surprise, the mathematics of probability seeks to identify the range of possible outcomes and to offer ways of measuring their likelihood. This is easier in some scenarios than others; when the baby is born, it will be either a girl or a boy!

Theoretical probability allows us to imagine what a perfect set of data would be like and is suited to certain types of event such as tossing coins or rolling dice where we assume the outcomes are equally likely. In a perfect world if I toss my coin ten times I will get five heads (H) and five tails (T); experimental probability, however, tells me that this may well not happen. Each and every time I toss the coin there is an equal chance it could be heads or tails so I could get any set of experimental data from every toss being heads through to all tails. The interesting feature of this is that the more experiments we do, the greater the likelihood that the experimental data will conform more closely to the theoretical probabilities (the perfect data set). Therefore probability benefits from using samples which are as large as possible. All probability is about uncertainty and mathematics seeks to quantify this.

Rationale for teaching probability

Probability is one of my favourite topics to teach. It is not always extensively represented in the primary curriculum, which is a great pity, as getting to grips with concepts involved in probability has clear mathematical and life benefits. Life is full of uncertainty and there are regular decisions to be made; these are informed by understanding the possible and probable outcomes. For young children this is about recognition of uncertainty from day to day, such as not knowing whether we will be able to go out to play this morning as it looks like it might pour with rain; we do not know for sure but we can make a simple prediction based on the evidence we can see (the big black clouds). Decision-making can also be modelled, with the teacher weighing up options and articulating arguments for and against. Estimation is also linked to probability to a certain extent; I get used to roughly what a certain number of items looks like and can therefore make a reasonable prediction as to how many I think there are. Similarly, tests for divisibility are supported by a sense of likelihood; this number is even so it must be divisible by 2 and there is a possibility it is divisible by 8.

As human beings we are exposed to probability through experiences related to gambling; anything from a raffle at the school fête to playing the tables in Las Vegas. This is not the place to discuss the rights or wrongs of such activities, but ensuring children develop an understanding of the mathematics of probability at least ensures they better understand their chance of winning. If I buy a lottery ticket then I have a chance of winning a lot of money, but it is a very slim chance! Incidentally, the first lesson for some young children can be somewhat more basic; I have seen little ones get excited by school raffles without realising you have to have a ticket to stand a chance of winning, whilst others may not appreciate that winning is by no means guaranteed, however many tickets you happen to have. Well, unless no one else bought any tickets, that is!

Good mathematical reasoning opportunities occur most readily in rich mathematical tasks; probability investigations offer some of the richest, as we will see in the activities section using dice and such like. Because those investigations generate data, children will be employing and rehearsing data handling skills to record and interrogate the information (see Chapter 8). Finally, because the measurement of likelihood involves numerical conventions, children will need to draw extensively on some of the number skills discussed in Chapter 2.

Progression in probability

From the language of probability to the numerical conventions

Given that children's earliest experiences of probability are everyday ones, it is no surprise that everyday language plays a big part to start with. Words and phrases like 'impossible', 'no chance' and 'certain', 'definite' describe the two ends of the spectrum. The words used in between are somewhat trickier to interpret and compare; something could be deemed 'fairly likely', 'probable' or 'quite likely' but as to which is more or less likely, that would be difficult to determine. Deciding the likelihood of events can be very subjective and is rather dependent on how much information is available to the person judging the likelihood of a particular thing happening.

At a later stage numerical conventions are introduced, and at this point comparing and ordering probabilities becomes easier. Whilst we do not stop using the language of probability when we are introduced to the numerical conventions associated with it, numbers can help to give us a more precise way of describing the measure of how likely something is to happen. Incidentally, older primary children may need the occasional reminder that when a probability answer is called for, the type of answer expected will tend to be a numerical one.

Common misconception

'When the test said "What's the probability of picking...?" I thought "very unlikely" was a good answer'

In many ways the child could be right as "very unlikely" could have been a perfectly valid response to whatever was being asked. Typically, however, it will be a numerical rather than a verbal answer that is required.

Specific terms relate to specific numbers. For example, the term 'even' and the phrase 'fifty-fifty' both describe an event which has a $\frac{1}{2}$, 0·5 or 50% chance of happening – something like the equal chance of getting heads or tails discussed earlier in relation to theoretical probability, or the measure of how likely it is that an even number will be divisible by 4. Numerical values are not always suitable of course; asked what the probability is that Doug will eat curry for dinner tonight could be tricky to put in terms of numbers and rather depends on Doug's penchant for curry and a number of other factors. But where numbers are suitable, they are calculated based on understanding the possible outcomes and typically described in terms of fractions, decimals and percentages. The portion of the number system we use is essentially only the bit from 0 to 1, with '0' representing the certainty that something won't occur and '1' (or 100%) the certainty that it will. This, if you like, is the measurement scale we use for probability.

Occasionally I meet people whose attitude towards fractions is that they are rather outdated in relation to decimals, but where probability is concerned they are perhaps our most useful ally. Picture a chest full of sweaters; let us imagine there are ten, all different colours, including a red one. Reaching in and pulling out a sweater at random, our chance of picking out the red one has a probability of 0·1 or $\frac{1}{10}$ or 10%. In words we might find ourselves saying that we have a one in ten chance of it being the red sweater. Now consider a chest containing only nine sweaters (still just the one red one). Stating that we have a $\frac{1}{9}$ chance of delving into the chest and pulling out the red one is straightforward; decimal and percentage answers are going to be considerably more challenging and awkward. There is, however, an exception; fractions can be harder to compare unless they have a common denominator. Therefore we might want to convert into decimals (or percentages) if we need to be able to compare probabilities. The different numerical conventions are put to the test in the self-test section at the end of the chapter.

Another term associated with probability is 'odds'. This is a way of referring to probabilities typically related to the return being offered by a bookmaker, for example, in horse racing. If you back a horse being offered at odds of 3 to 1 and it wins, you get your stake money back plus three times the amount. So if you bet £2 and the horse wins, £8 would be paid out (£2 × 3 + £2). This would be a horse with fairly 'short odds'; one which the experts think has a good chance of winning. Sometimes the odds are 'odds on'; this is where the first number is lower than the second and suggests the horse is deemed to have a better than even chance of winning. So with odds of 1 to 2 ('2 to 1 on' in bookmaker-speak), and assuming the horse does win, your £2 bet would return only £3; the £2 stake plus half again. A horse with 'long odds', such as 50 to 1, might be attractive as you would get a good return on your money if it won; but the odds suggest it will not! Much like the insurance companies, the bookmakers' main aim is to make a profit and the odds offered across the whole race reflect this.

Approaches to recording probability

Identifying different outcomes in any given situation is key to understanding probability, as already discussed. The simplest way for young children is simply to list or draw the various outcomes as in Figure 9.1, where the child has worked out that there are four possible outcomes if I toss two coins. Where more outcomes are possible, children will need to work systematically to ensure they have found all the possibilities; see, for instance, activity 9.15 for quite a challenging example of this.

I tend to use 2p and 10p coins with children. They are a good size for easy handling but very obviously different as one is a copper coin and the other silver.

Figure 9.1 Possible outcomes when tossing two coins

Common misconception

'If I toss two coins, there are three possible outcomes (HH, TT, HT) and a $\frac{1}{3}$ probability of each'

You will have noted in the example in Figure 9.1 that the two coins depicted were different ones; this was very much intentional. Using identical coins makes it very difficult to convince children that HT is different from TH!

Recognising different outcomes such as HT and TH can be a major difficulty and should really come as no surprise. We spend years hoping children will realise that 2 + 3 and 3 + 2 are essentially the same and then we expect them to recognise that they are different in the context of probability. Activity 9.6 describes a game of bingo using two dice, and a total of 11 is theoretically easier to roll than 12. Whilst a total of 12 has to be 6 + 6, a total of 11 can be 5 + 6 or 6 + 5. If your children struggle with this, they are not alone, but it may help to use two different coloured dice; as in the coin scenario, this may help to make the difference between the two outcomes easier to distinguish.

Tree diagrams offer a way of arriving at these possible outcomes in an organised fashion, though it should be noted that this approach becomes unmanageable beyond a certain point. Imagine an investigation involving tossing a coin and rolling a dice. The tree diagram in Figure 9.2 helps us to see the possible outcomes and to identify how probable it is that we will get a certain result. There are 12 separate outcomes, therefore each one has a $\frac{1}{12}$ chance, but if we wanted to get heads and an odd number, we can see there would be a $\frac{1}{4}$ chance since three out of the 12 answers would satisfy this.

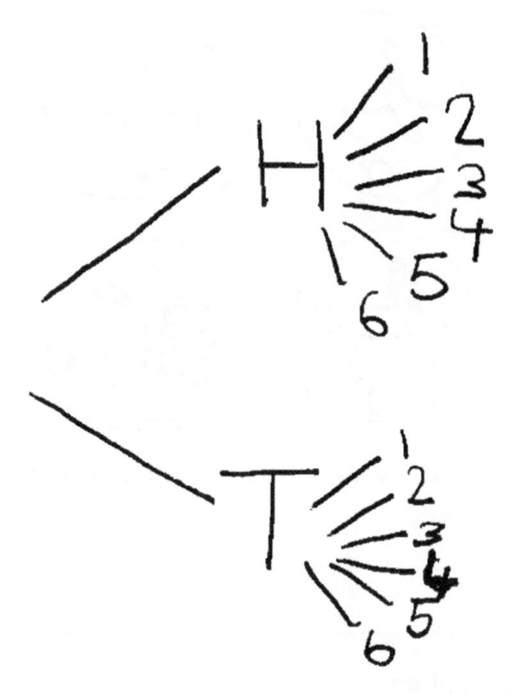

Figure 9.2 Tree diagram showing the possible outcomes of tossing a coin and rolling a dice

Two-way tables provide an alternative to tree diagrams and work well where there are two independent events. The same outcomes, the results of the coin tossing and dice rolling, can be seen in Figure 9.3 in a two-way table.

Later in their education children will be introduced to the idea of identifying numbers of outcomes in ways other than using diagrams or working systematically.

	1	2	3	4	5	6
H	H1	H2	H3	H4	H5	H6
T	T1	T2	T3	T4	T5	T6

Figure 9.3 Two-way table

Remaining with our coin tossing and dice rolling example, if I wanted to work out the chance of getting tails and a five I would multiply the two probabilities; there is a $\frac{1}{2}$ chance of getting tails and a $\frac{1}{6}$ chance of rolling a five, so the overall probability is $\frac{1}{12}$ ($\frac{1}{2} \times \frac{1}{6}$), as we can identify from either diagram. Note that the sum of all possible outcomes will always be 1; so if I have a $\frac{1}{12}$ chance of getting a five and tails, I have an $^{11}/_{12}$ chance of *not* getting that combination.

Key terms describing aspects of probability

The activities that generate data in the field of probability are often referred to as 'events'; theoretical probability allows us to predict likely outcomes of events, whilst experimental probability relates to raw data. There are two main terms which help to describe such events. When we toss a coin, the outcome is only ever going to be heads or tails, and these outcomes can be described as 'mutually exclusive'. This means that both cannot occur at the same time; the outcome will be *either* heads *or* tails. Each successive toss of a coin is also an 'independent event'. The occurrence of heads, say, makes it no more or less likely that the outcome of the next event will be heads. Each individual event is independent of those before and after it and there is an equal chance every time of the result being heads or tails.

Let us now consider picking successive cards at random from a pack and not replacing them; these are actions which may or may not prove mutually exclusive, but which will not be independent. The lack of independence occurs because the choice of the first card will influence the probability of the next choice. Imagine that we hope to get a red card on our first go; we have a 50% chance of getting lucky since 26 of the 52 cards are red ones. Now let us assume that we did get a red card and are about to choose our second card, hoping this time that it will be black. What are our chances? Well, better than they would have been for a red card. As one red has already been drawn, there are now only 25 reds out of 51 cards in total, therefore our chance of drawing a black card is marginally higher at $\frac{26}{51}$. Note that if the card is returned to the pack each time and the pack is well shuffled, the picking of a card becomes an independent event again. Each of the 52 cards are present in the pack every time and you have a $\frac{1}{52}$ chance of picking a particular card, even the one you have just put back!

Some aspects of the playing cards are mutually exclusive; we cannot get hearts at the same time as getting spades; we cannot get a three if we get a king; clubs are not red cards. However, other outcomes can occur at the same time; a red king or a black three are both possibilities where one category does not exclude the other (combinations of colour and picture, or colour and number in these examples). Specific cards essentially work in the same way; for example, a card can be red, a diamond and a queen all at the same time.

Combined independent events such as tossing a coin and rolling a dice can safely be considered in any order, both from a practical point of view and how the events are recorded. The child's tree diagram and the two-way table in Figures 9.2 and 9.3 both show the coin outcomes first, but it would not have mattered had it been the dice recorded first (3H rather than H3) as the two events are independent of each other. So from a practical perspective independent events such as coin tossing and dice rolling can occur simultaneously or sequentially; it will not make any difference to the results. This would be the same for more than two independent events, although the recording of the numerous possible outcomes would need to be reconsidered.

Activities for teaching associated with probability

Discussion and questioning will be key to ensuring that many of the activities in this section work well with your class. Most of the activities can be successfully used with a variety of age groups and adapted to place the emphasis on different aspects of probability; but most of all, enjoy using them!

Note that probability activities do not tend to come with a guarantee. What we hope an investigation might demonstrate, it might not, precisely because we cannot predict what will happen – one of the lessons we are setting out to learn! Thus we might reasonably expect 100 tosses of a coin to give us a good balance of heads and tails, but we have to accept that it might not.

Pupil activity 9.1

Picture book probability
Y1 to Y6

Learning objective: To understand that predicting what happens next in a story relates to what is *likely* to happen next.

A key facet of probability is prediction of likely outcomes, and stories offer a way of exploring this with young children, although the activities would also be beneficial with older children and have strong links with literacy. Few books offer explicit probability content, yet most can be used to focus on the idea of likely outcomes. Two possibilities are outlined below:

- Take a story which is already very familiar to the children and consider alternative outcomes. For older children this could develop into writing stories with optional endings where a character's decision en route will influence what happens!

- Select a book which you are pretty sure will be unfamiliar to the children. Stop reading part way through. What do they predict is likely to happen next? What was their rationale for suggesting it? How likely is their idea in comparison with the ideas of other children? If they had to order the ideas from least likely to most likely, which order would they put them in and why?

When you next read to your class, consider the questions you ask your children. Any questions which relate to what comes next, or perhaps a decision that the central character needs to make, are probability-related. A simple example is *The Odd Egg* (Gravett, 2008) where the children might guess what the unusual egg could hatch into. This suggests the possibility of links with science, where eggs of particular colours, shapes and sizes are likely to hatch into certain types of bird and animal.

Another huge collection of stories written for very young children, all of which start *That's not my* ... (published by Usborne Children's Books) lend themselves to the idea of sorting according to property. The fact that 'that's not my monkey; its tail is too velvety' and 'its tongue is too fuzzy' can be used to illustrate aspects of probability through identifying the most pertinent characteristics and using these to decide whether the item could be the one you are seeking.

Pupil activity 9.2

Probability problems
Y1 to Y6

Learning objective: To develop understanding of likelihood.

Many everyday scenarios lend themselves to setting problems with a probability context. Consider the examples below, set in the context of a school car park:

- In the car park this morning I noticed that there were only red and black cars: 7 black ones and 10 red ones. What's the most probable colour to be driven from the car park first? How likely is it?
- What's the probability that the next car to drive in will also be red or black? [This is the sort of question with no easy answer and therefore ideal for promoting discussion. Having a car park full of red and black cars does not influence the colour of the next one to drive in; it would be better to ask about statistics relating to the colour of cars typically owned by drivers. And if I wanted to be really particular, I could question how this data was gathered, whether it might be regional, whether there is any difference between men's and women's choices, and so on – after all, there are more female primary school teachers!]

Pupil activity 9.3

What's in the bag?
Y1 to Y5

Learning objective: To develop and refine understanding of likelihood.

Preparation for this activity requires you to secretly put a selection of Multilink into a small bag you cannot see into so the children do not know the number or colour of cubes the bag contains. Ideally you will put in considerably more of one colour than any other. (I often go for ten red ones plus one of each of every other colour.) The children are then given the chance to dip their hand into the bag without looking but first have to predict what colour cube they think they will pull out, which therefore assumes they are familiar enough with Multilink to know at least some of the colours which are possible. The colour is announced or shown to the class before being put back in. As the game progresses children's choices of colour begin to be influenced by what they have already seen; for example, getting an inkling that there might be more of one colour, or trying for a colour they have not yet seen.

Once lots of cubes (ideally just short of 20) have been seen, explain to the children that there are 20 cubes in the bag and ask them (in pairs, as this will promote discussion) to predict how many of each colour they think there might be. Share some pairs' ideas and ask the children to articulate what made them opt for the colours and ratios they did. Note that the children may or may not recognise that all 20 cubes cannot have had an outing; even if more than 20 have been chosen, this is still likely to be the case. There may therefore be some colours that we have not seen; do the children suggest any such colours as part of their prediction or do they limit their choices to the ones they have already seen? Eventually they can be allowed to see the contents of the bag, which will include surprises for some. Time should be allowed for discussion.

Following this type of game, orchestrated by the teacher, the children can be given opportunities to make up their own and to use the numerical conventions of probability to describe the likelihood of picking out certain colours. Do experiments produce data which bears this out in practice?

Pupil activity 9.4

Expected outcomes
Y3 to Y6

Learning objective: To appreciate that generating more experimental data may lead to probabilities closer to expected results.

This activity revolves around collecting experimental data about whether you get heads or tails when spinning or tossing a coin ten times. Ask the children to predict what they think the outcome will be before they try it out in pairs (working with a partner makes it easier to undertake and record the results, perhaps using tallying). Since few results are likely to conform exactly to the five–five split that some are bound to have predicted, discuss why the children think this is the case. Then combine all the results from the pairs in the class. Hopefully doing so will provide data which matches the theoretical probability more closely – but there are no guarantees!

Pupil activity 9.5

Tossing two coins
Y4 to Y6

Learning objective: Ability to distinguish between similar outcomes.

This activity seeks to overcome the misconception outlined earlier in the chapter that the probability of getting HH, TT or HT with two coins would be $\frac{1}{3}$ apiece, noting that the same misconceptions may or may not exist to start with in your own class. Give pairs of children two different coins and ask them to predict what they think the probabilities are for getting HH, TT and HT. Do not at this stage attempt to influence their views, just give them time to discuss their ideas with their partner and to record their prediction. The prediction should then be put to the test experimentally, at the end of which the children may wish to revise their probabilities. Asking those who have changed their minds to articulate why may support others in being sure that the probabilities are actually $\frac{1}{4}$ (HH), $\frac{1}{4}$ (TT) and $\frac{1}{2}$ (HT or TH), though it may be necessary to be very explicit about the fact that HT and TH are different by looking closely at the fact that the coins are the other way round.

Developing awareness could equally be related to rolling two dice, making sure that these are different colours to better illustrate the concept.

Pupil activity 9.6

Probability bingo
Y1 to Y6

Learning objective: To appreciate possible outcomes and begin to realise they are not all equally likely.

Regular bingo is all about the luck of which numbers come up, and this activity builds on that. The children's 3 × 3 bingo boards are filled by themselves based on the totals they think they will get when two dice are rolled and the numbers added. It needs to be made clear that they can use numbers more than once but can only cross out one at a time. I have known children to put anything from 1 to 12 in their grid and it usually dawns on them after one game that they're never going to win with a 1 in the grid! Playing quite a few times without discussion of any mathematics is beneficial; this gives children time to realise that some numbers seem to come up more than others and that some numbers can be hard to cross out. This means that each time they play, they may make different choices about the numbers to include in their grid, and of course they cannot have them all anyway as there are only nine spaces, so in its own way the size of the grid helps to force the issue.

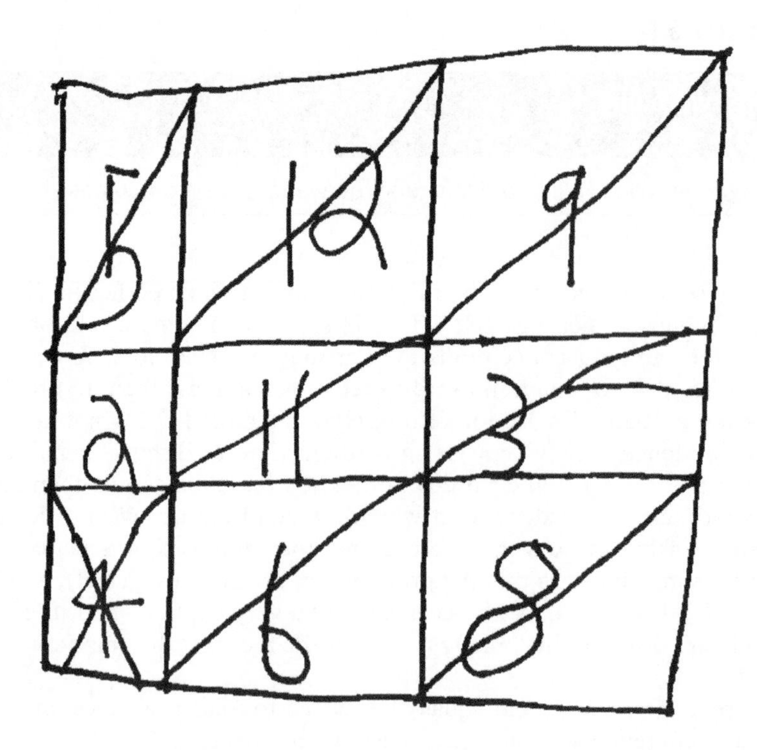

Working out why some numbers occur more than others is slightly more challenging; encourage the children to discuss their theories and to think about it over quite a long period of time. This need not be in maths lessons; mulling it over as 'homework' will give the children time to reflect (something I suspect we do not do enough in our mathematics teaching). Two-way tables then offer a more formal way of illustrating the possible outcomes and here we can see why 7 is a more likely total than any other. Does that mean I would only put sevens in my grid? Well, no, since some of the other numbers have almost as much chance of occurring. Remember that two colours of dice may aid children's appreciation that results like 2 + 3 and 3 + 2 are different.

+	1	2	3	4	5	6
1	2	3	4	5	6	7
2	3	4	5	6	7	8
3	4	5	6	7	8	9
4	5	6	7	8	9	10
5	6	7	8	9	10	11
6	7	8	9	10	11	12

Do not forget that chance will play a part in this activity; whilst you will theoretically find it harder to throw 12 than 7, you cannot be sure what will happen in practice!

The game can of course be adapted with the numbers multiplied instead of added, with the most and least likely totals apparent in a two-way table.

Younger children can play in a way more akin to real bingo, deciding which four numbers to put in a 2 × 2 grid and crossing a number out each time it is rolled. Here the mathematics of probability is not really involved, but children will still have the opportunity to realise that you cannot be sure your number will come up; it's luck!

Pupil activity 9.7

Spinners
Y1 to Y6

Learning objective: Ability to calculate theoretical probabilities.

Rather than generating experimental data, this activity focuses on theoretical data by exploring what the likely probabilities would be, based on spinners. Key questions for the youngest children relate to numbers it's impossible to get on each the spinner as opposed to the numbers we can get. The spinners don't have to show numbers incidentally; the sections could show pictures of animals, or whatever you like.

As understanding progresses, children can begin to identify the likelihood or theoretical probability of each of the possible outcomes. Note that these will always add up to 1.

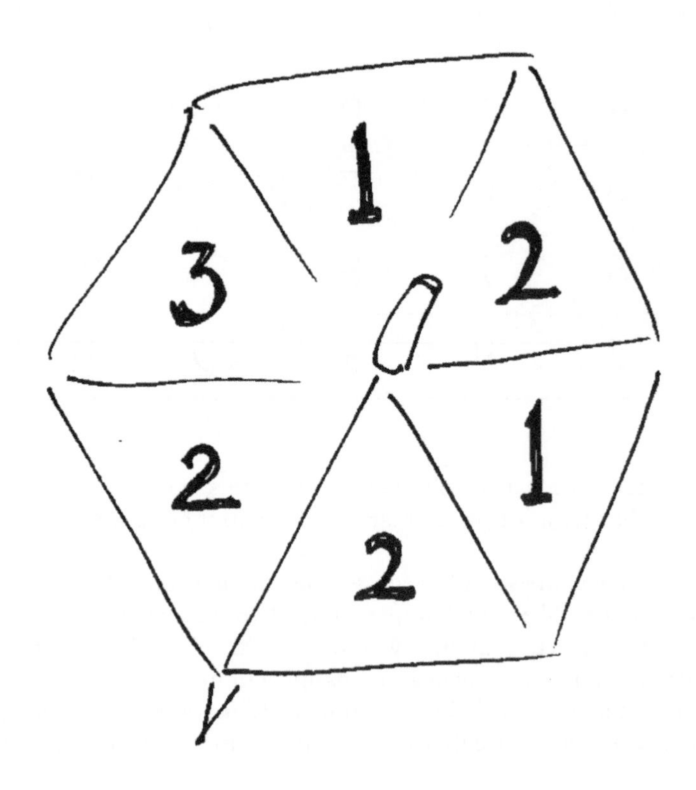

This example spinner has been chosen because of the possible misconception:

Common misconception

'There are three different numbers on my spinner so each number has a $\frac{1}{3}$ probability'

The fact that there are different numbers of each makes a difference of course. In the case of the spinner we see here we have an even chance of getting a 2, given that half the numbers on the spinner are twos.

Various spinners are available for use on the interactive whiteboard and these are sometimes displayed in pairs. Using both in tandem makes a really nice extension for the most able. Looking at the ones pictured here, what totals could we make by spinning both and adding the two numbers? How can you be sure you have found all the possible totals? Of all the totals we could make, which would be the most likely and why? Describe the probabilities for each total.

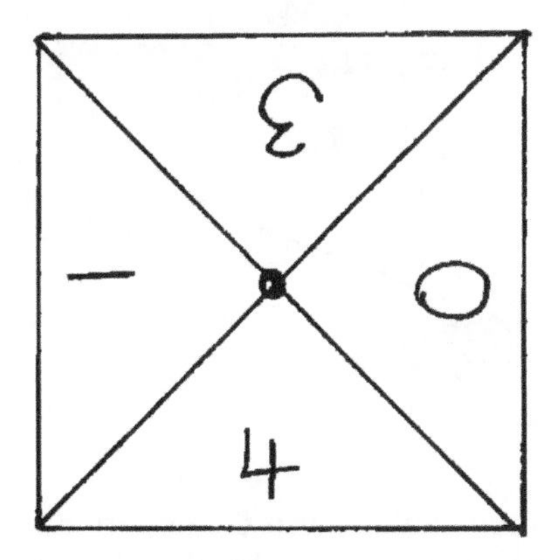

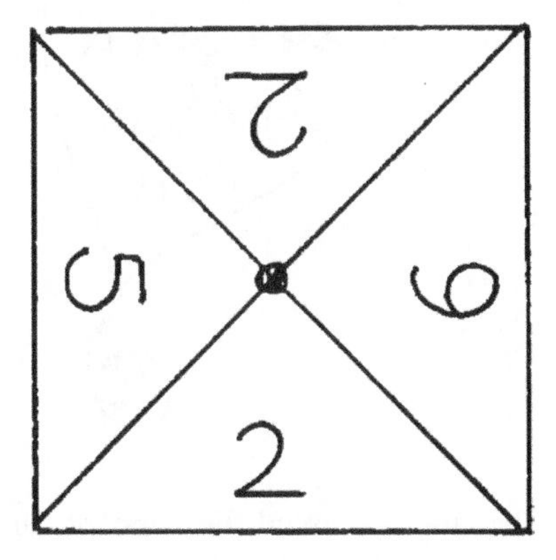

Rather like the bingo in activity 9.6 where the dice total outcomes were identified in a two-way table, the same approach could be encouraged here, particularly if children are struggling to find the outcomes in other ways.

Pupil activity 9.8

Agree or disagree?
Y3 to Y5

Learning objective: To appreciate that events can sometimes be mutually exclusive.

Whilst children of this age would not be expected to understand the language of mutual exclusivity, we can certainly begin to introduce them to the concept.

Having discussed the concept cartoon, children can be given the opportunity to explore this idea through other resources, looking for outcomes which can both happen at the same time and those which cannot. An everyday example would be something like considering whether it can rain and be sunny at the same time.

Pupil activity 9.9

Tree diagrams: are they both right?
Y4 to Y6

Learning objective: Ability to understand the independence of events.

Again, rather than the actual term 'independent event' children are really being introduced to the concept. Earlier in the chapter we met the idea that tree diagrams depicting two events such as coin tossing and dice rolling can be drawn in either order; studying some examples with the children may help to illustrate the nature of independent events and clarify the children's thinking regarding whether the activities have to take place in a particular order, or even simultaneously.

Pupil activity 9.10

Different games
Y5 to Y6

Learning objective: To begin to appreciate how changing a game can affect the probability of an event.

Building on the children's understanding of the concept of independent events, this activity encourages them to think about how the rules for different games can mean that outcomes are no longer independent. Compare the following games based on a bag with 9 marbles: 2 red, 3 blue, 4 yellow:

- The aim of the game is to pick a blue marble from the bag. The marbles are returned to the bag after each go. What's the probability of succeeding? [This has the same probability each time; each outcome is from an independent event, so you have a $\frac{3}{9}$ or $\frac{1}{3}$ chance.]
- The aim of the game is to get a blue marble on the second go; the first marble is not returned to the bag. What's the probability of this? [These outcomes are no longer independent; depending on what colour marble we pick the first time, the chance of our second marble being blue will be better or worse.]

The probabilities involved in the second game are quite complex; for primary-age children it will be sufficient to realise that the first marble picked affects what we hope to get the second time.

Pupil activity 9.11

Understanding a pack of cards
Y5 to Y6

Learning objective: Ability to state probabilities relating to numbers of playing cards as fractions.

Start by exploring the events that would be certain in picking a card from a pack of 52 cards; for example, that the card will be red or black, that it will not be purple, that it will belong to one of the four suits. Having banned these sorts of category, the children can take it in turns to decide the card they hope to pick at random from the pack; for example, 'I hope to get a club and I have a $\frac{13}{52}$ or $\frac{1}{4}$ chance.'

The cards can either be returned to the pack each time since this keeps the probabilities the same, or can be displayed clearly for all to see so that calculating each probability can take into account the cards which have already been drawn. For example, if the first card is the queen of clubs and the next player hopes to get a clubs card, this would now have a probability of $\frac{12}{51}$ since there is one less card and it was from the same suit.

Choices can revolve around colour, suit, number (specific, odd/even or just 'a number card') or picture card (specific choice, or any picture card) either singly or in combination. For example, 'I hope to get a club and I want it to be a number card. This has a probability of $\frac{10}{52}$.'

Pupil activity 9.12

Probabilities on the number line
Y2 to Y6

Learning objective: To develop ability to compare probability terms and numbers.

A washing line-style number line works well for probability since words and values can be so easily moved around once they have been pegged up. Young children can work to discuss and position words from 'impossible' to 'certain', whereas the older children can decide collaboratively where a range of numerical values belong. The fractions generated in activity 9.11 using playing cards would provide a nice extension since the values generated here will be more challenging to position than, say, halves and quarters, making a nice platform for some quite advanced number work.

Working with words and phrases ensure that you sometimes include some contentious ones; for example, using lots of examples with the same root word qualified in different ways:

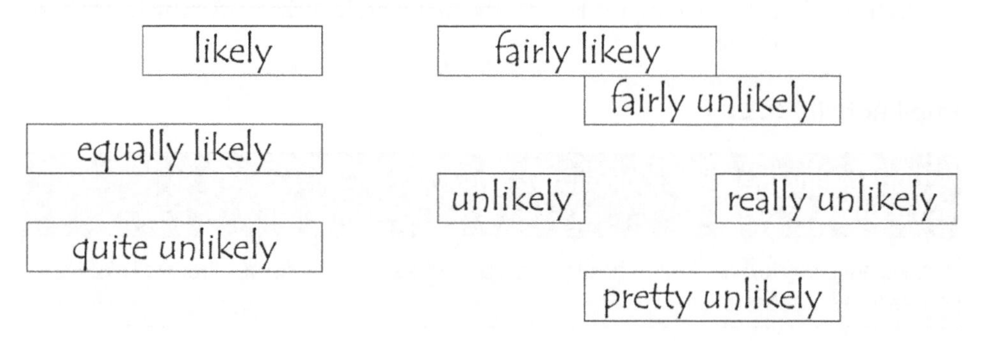

These might be quite hard words and phrases to decide where to peg on the line, but they reflect linguistic variety realistically and could be used for a rough comparison with their numerical counterparts. Sayings related to probability may also be familiar to some children and these would be equally difficult to position: what does somebody mean, for example, when they say 'the chance would be a fine thing'?

Pupil activity 9.13

Ordering of events
Y1 to Y6

Learning objective: To develop understanding of statistical likelihood.

The simplest version of this activity is to give children a selection of statements, based on everyday scenarios, to put into order according to which they think are most or least likely to occur. Children can of course make up their own statements.

An alternative way of approaching this is to give little titbits of information in separate envelopes to be opened in a set order. This way the children can be forced to change their minds as more information comes to light:

1. Dad really likes motorbikes. What's the probability he will ride one on Sunday?
2. Dad doesn't own a motorbike any more. What's the probability he will ride one on Sunday?
3. Dad is visiting his friend on Sunday and his friend has two classic motorbikes. What's the probability he will ride one on Sunday?
4. The weather forecast for the weekend is awful. What's the probability Dad will ride a motorbike on Sunday?

What other scenarios might give rise to a similar series of probability questions? Once the children's ideas are gathered in language terms, they could be asked to give events a likelihood score out of 100. This is as a way of moving from rough ordering to giving the probability using some sort of number scale.

Pupil activity 9.14

Revisiting the number system
Y1 to Y6

Learning objective: To appreciate risk in the context of probability.

Activities 2.3 (ladder game) and 2.6 (digit decisions) are described in Chapter 2 as they support children's understanding of the number system, but in both cases they also give children a chance to experience probability. Just playing the games, the children will begin to appreciate that it can be difficult to decide where to place the number or the digit as the games involve uncertainty. The risks, whilst they cannot be eliminated, can be managed by thinking carefully about what is likely. As the children progress, place greater emphasis on their ability to articulate why they choose to put the numbers where they do. With the oldest children this can be extended to calculation of actual probabilities to influence the decisions.

Pupil activity 9.15

Human fruit machine
Y5 to Y6

Learning objective: Ability to identify all outcomes in order to judge theoretical probability.

This activity is one I first saw at a large Guide camp many years ago, where it was being used as a fundraiser. It is ideal for use at school fêtes since the children can easily run it themselves and preparation can involve extensive probability work! Three children each hold a bag containing three pieces of fruit, for example, one banana, one apple and one pear each. Once a customer has paid to play, he or she counts to three; on three each child pulls a fruit out of their bag; if all three are holding the same the customer wins a prize. But how often can we expect someone to win?

In order to identify the likelihood of getting all three the same (always assuming it is totally random, with the children not collaborating in any way), secondary children might well use an algorithmic approach. With primary children it would be more appropriate to explore all the possible outcomes by working systematically. This could be approached by making a tree diagram, but the child here has listed the various combinations, determining that $\frac{3}{27}$ or $\frac{1}{9}$ will be winning combinations.

We have always worked from the premise that it is better to charge less and have more people visit the stall and enjoy playing, but knowledge of the probabilities will help you and the children to make informed decisions about how much you can afford to

AAA
AAP APA PAA
AAB ABA BAA
BBB
BBA BAB ABB
BBP BPB PBB
PPP
PPA PAP APP
PPB PBP BPP
ABP
APB
BAP
BPA
PBA
PAB

spend on prizes in relation to how much you will charge people to play. Clearly you want to take more money than you have to outlay!

Summary of key learning points:

- probability is based on trying to predict outcomes and can be theoretical, experimental or statistical;
- outcomes are typically uncertain and may or may not conform to expected probabilities;
- understanding probability ensures that we can make informed decisions;
- both language and numerical conventions can be used to describe probabilities.

Self-test

Question 1

I am going to pluck a flower from a vase containing four tulips and five daffodils. Assuming I do not look, what is the chance that my flower will be a tulip? (a) $\frac{4}{9}$, (b) 44%, (c) 0·444

Question 2

There are three marbles in a bag: one red (R), one blue (B) and one green (G). If I take one out of the bag, followed by another, without returning the first one to the bag, what is the probability that the second one I pick out will be red? (a) $\frac{1}{9}$, (b) $\frac{1}{6}$, (c) $\frac{1}{3}$

Question 3

Playing the card game described in activity 9.11 generated the following fractions: $\frac{13}{51}$ and $\frac{12}{49}$. If I put these fractions on a number line, $\frac{13}{51}$ will come closer to the certain end (with a value of 1) than will $\frac{12}{49}$. True or false?

Question 4

Tossing three coins, I am hoping to get HHH. The probability of this is (a) $\frac{1}{8}$, (b) $\frac{1}{9}$, (c) $\frac{1}{4}$

Self-test answers

Q1: (a, b, c) All answers are actually pretty much equivalent, but the fraction (a) provides our easiest and most accurate answer. Both the percentage and the decimal answer have been rounded since ninths give rise to recurring decimals.

Q2: (c) These are not independent events since the colour of the first marble affects what is left in the bag. Drawing a tree diagram suggests that there are six possible outcomes, each of which depend on the first draw: RB, RG, BR, BG, GR, GB. Of these, two have red as the second marble, therefore there is a $\frac{2}{6}$ or $\frac{1}{3}$ chance that the marble picked out second will be red.

Q3: (True) Using a calculator is the easiest way to turn the fractions into decimals by dividing the numerator by the denominator in each case. This makes the values much easier to compare. We find that $\frac{13}{51}$ is 0·255 (to three significant figures) whereas $\frac{12}{49}$ rounds to 0·245, so $\frac{13}{51}$ is just marginally closer to 1.

Q4: (a) $\frac{1}{8} = \frac{1}{2} \times \frac{1}{2} \times \frac{1}{2}$, with $\frac{1}{2}$ being the probability that each time I will get heads rather than tails. Another way of approaching this is to consider all possibilities: There are eight in total (HHH, HHT, HTH, THH, HTT, THT, TTH, TTT) and only one of them is the one we are looking for, therefore the probability is $\frac{1}{8}$.

Misconceptions

'When the test said "What's the probability of picking...?" I thought "very unlikely" was a good answer'

'If I toss two coins, there are three possible outcomes (HH, TT, HT) and a $\frac{1}{3}$ probability of each'

'There are three different numbers on my spinner so each number has a $\frac{1}{3}$ probability'

10
Shape

About this chapter

This chapter details various issues associated with the teaching of two-dimensional (2D) and three-dimensional (3D) shape; children in classes with practitioners aware of such issues are more likely to establish solid foundations regarding shape concepts. Essential subject knowledge is explored under the following headings:

- 2D and 3D shape families and their properties
  - Illustrating variety within shape families
  - Decoy shapes
  - A focus on four-sided shapes
  - Regular and irregular shapes
  - Shape orientation
- The language of shape
  - Mathematical terms to avoid ambiguity
  - Congruence and similarity
- Tessellation and tiling

Symmetry, whilst a related topic, is covered in the next chapter.

What the teacher and children need to know and understand

2D and 3D shape families and their properties

One of the most important factors underpinning the teaching of shape is the appreciation that shapes belong to different families because of their properties, and that these families often overlap with each other. As we progress through the chapter, be on the lookout for any misconceptions you or your children might have in relation to this. Mathematicians like to be concise, and it helps in the long run if we can encourage children to think about what information is sufficient when defining particular shapes by their properties. There are some properties that are essential, and contrasting these with features that can be altered is a useful way of looking at this. For example, a triangle must have three corners joined by straight lines, but the sizes of the angles and the lengths of the sides can vary, as can the overall size and also colour. Get into the habit of drawing the children's attention to the essential versus the transient aspects.

Illustrating variety within shape families

In order for children to become conversant with shape families, one of our main roles when teaching shape is to ensure that the range of examples children are presented with is really rich. This variety helps to ensure children develop a broad understanding of the shape in question rather than a rather narrow, limited concept.

With regard to 2D shape we know that people asked to draw a triangle will often produce an equilateral triangle; sadly, children who regularly experience only triangles of this type, fail to recognise other types of triangle.

> **Common misconception**
>
> **'All triangles look the same'**
>
> So whatever resources you choose to use for the teaching of 2D shape, from those you draw yourself, to the plastic shapes you can buy, make a conscious effort to choose and present variety.

For a diverse range of 3D shapes, junk boxes are an absolutely fabulous resource. Bought collections (plastic, wooden) often have single examples of a type of shape so children form rather limited mental images, whereas what we really want is for children to recognise all cylinders, for example, by their key properties. Cylinders have identical circular ends, parallel to each other and joined by a curved face – the essential features. But cylinders can be tall and thin or short and fat, they can be made of all sorts of different materials, be different colours – all transient rather than essential features.

Figure 10.1 Variety of cylinders

Whilst the solids in Figure 10.1 appear very different, children need to come to recognise all of them as cylinders and thus to develop a rich appreciation of what constitutes a cylinder. Cylinders have some similarity to members of the prism family (end faces are parallel and the same), but they are not conventionally considered to be prisms since cylinders include a curved as well as plane (flat) faces whereas prisms have only flat faces. A good variety of triangular prisms can often be found in sweet shops, though these tend to involve equilateral triangles (all three edges and angles the same). Triangular prisms with non-equilateral triangles are harder to come by, but even if you cannot find any, try to at least draw children's attention to the possibility. Watch out for other prism examples, named by their end shape, to add to your collection (pentagonal prisms, hexagonal, etc.). Prisms are also recognisable by the rectangles that join the 2D shapes at either end; if you find the two end faces are offset and joined by triangles rather than rectangles, then it is possible you have stumbled upon an anti-prism!

Decoy shapes

An aspect of teaching in shape which seems to be too often overlooked is the opportunity for children to distinguish between actual examples of shapes and what I am going to refer to as 'decoy shapes' – those 'almost but not quite' examples. To prove they have really solid understanding regarding the properties of a particular shape, children have to be able to spot those which do not quite fit the category. This has implications for your choice of examples: select some which at first glance may appear to be OK, but which on closer inspection have too many sides, or curved sides, or not quite the right type of faces and so on. Figure 10.2 focuses on triangles, but different choices will allow you to focus on other 2D and 3D shapes equally well.

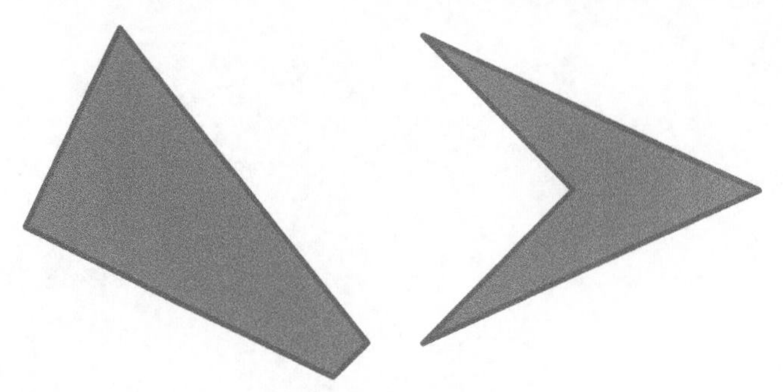

Figure 10.2 A couple of 'decoy' triangles

An opportunity to explore this in the context of 3D shape could be to show the children a typical sweet box like the one we see in Figure 10.3 and to ask whether or not the shape is a cuboid; after all, it has the same number of faces (six), the faces are quadrilaterals – with cuboids all six faces have to be rectangular, however, so this cannot be a cuboid. In case you are wondering, these are referred to as 'truncated pyramids', and the same term can be used to describe something like a plastic cup – a truncated cone.

Figure 10.3 A tasty truncated pyramid!

A focus on four-sided shapes

With 2D shape, squares typically seem to cause the greatest conceptual difficulty for adults, quite possibly because they are members of so many different shape families, but have a special name 'square' used from a young age. Over time the term 'square' seems to stick, with learners reluctant to call the shape by any other name!

Common misconception

'A square is not a rectangle'

In order to appreciate some of the other shape families claiming 'hey, square, you're one of us!' let's take each family in turn and explore what properties would be required to be a member.

2D shape	In the early stages a 2D shape may be referred to as a flat shape. Are squares flat? Yes! So a square is a 2D shape.
Polygon	The term 'polygon' describes a 2D shape with only straight sides, so excludes shapes like circles. Do squares have all straight sides? Yes! So a square is a polygon.
Quadrilateral	This term describes any polygon with four sides (and, given the definition of polygon, the sides have to be straight of course). Do squares have four sides? Yes! So a square is a quadrilateral.
Rectangle	A rectangle is a quadrilateral with four right angles. Do squares have four right angles? Yes! So a square is a rectangle.
Parallelogram	Parallelograms are quadrilaterals with opposite pairs of parallel, equal length sides. We know a square is a quadrilateral, but do they have pairs of parallel, equal length sides? Yes! So a square is a parallelogram.

Figure 10.4 Families of shapes the square belongs to

Squares are sometimes referred to as 'special' cases of the shape families referred to in Figure 10.4; for example, a square's a 'special' rectangle because it has the required right angles but also has all equal length sides. In a similar vein, the term 'rhombus' describes a special type of parallelogram with equal length sides. By this definition, is a square a rhombus? Yes! Many of you will also be familiar with the term 'oblong'; this is typically used to describe rectangles that are not squares, in other words, rectangles where adjacent sides are different lengths. Activities 10.11 and 10.12 help us to deal with issues of shape family overlap.

The 'trapezium' and 'kite' families also relate to four-sided shapes but typical definitions preclude squares, though it should be noted that the trapezium varies in interpretation between countries. I usually work from the premise that: a trapezium is a quadrilateral with one, and only one, pair of parallel sides (Figure 10.5); a kite has adjacent pairs of equal length sides. I have always assumed the lengths vary from pair to pair.

Figure 10.5 Trapezium shown as the roof of a house, and a kite-shaped kite!

Regular and irregular shapes

Integral to the rich variety of shapes you will want to present the children with will be both regular and irregular versions of shapes. Awareness of irregular shapes and their incorporation in your teaching are really important. As with the triangle earlier, if all the children ever see are regular pentagons (five-sided 2D shapes), then when they meet more unusual examples they may not recognise them as members of the pentagon family because of the rather limited concept formed through experience. Note that opportunities to engage with irregular 2D shapes can be somewhat limited if you use ready-made worksheets, text books and internet sources.

Common misconception

'There are lots of regular quadrilaterals'

Because 'regular' implies all sides and angles will be equal, the square is our only example of a regular quadrilateral. Learners regularly try to include other examples, possibly because of their everyday sense of the term 'regular' with shapes like these seeming to fit the criteria!

As regularity or irregularity is not confined to lengths of sides, but includes size of angle as well, you will want to illustrate all the different types of angle:

- Acute (less than 90°)
- Right (90° exactly)
- Obtuse (more than 90° but less than 180°)
- Reflex (more than 180°)
- A straight line is effectively an angle of 180°!

Reflex angles become possible once we get to numbers of sides above three; a triangle cannot have a reflex angle on the inside! Any triangle's three angles add to 180°, and an equilateral triangle therefore has 60° corners. Other types of triangle, such as an isosceles triangle with two sides and angles equal, are mathematically irregular (however symmetrical they might look!). A triangle with all three lengths and angles of different sizes is referred to as a 'scalene' triangle, and both isosceles and scalene triangles can have right angles. Whereas the interior angles of a triangle always add to 180°, those of any quadrilateral add to 360°, any pentagon 540°, hexagons 720°,... (you're noticing a +180° pattern here, I'm sure).

Common misconception

'Right angles always look like an "L" shape'

This misconception arises because right angles are often shown in isolation in this orientation.

The main polygons children will meet beyond quadrilaterals are pentagons, hexagons (6 sides), heptagons (7 sides), octagons (8 sides), nonagons (9 sides) and decagons (10 sides); the internet can help you out if you need others. A 22-sided shape once came up in relation to Activity 10.17, and various internet sources suggested we could call it an icosikaidigon! As all polygons are made up of only straight lines, shapes like circles do not belong, instead forming part of what is referred to as the family of curves, although some define a circle as a polygon with an infinite number of sides.

It may surprise you to know that whereas we can go on and on naming regular polygons, with the 3D equivalent of regular shapes there are only five and these are referred to as 'Platonic solids'. The smallest in the set is the tetrahedron or triangular-based pyramid, consisting of just four equilateral triangles, with three joining around each vertex. Taking a look at the vertices is important as it is not only that all the faces are identical on a Platonic solid, each vertex should also be exactly the same. The remainder of the set are the cube (6 square faces, 3 joining at each vertex), the octahedron (8 equilateral triangles, 4 at a vertex), the dodecahedron (12 pentagons, 3 at each vertex) and the icosahedron (20 equilateral triangles, 5 around each vertex). Children should also meet irregular 3D shapes of course, and will easily do so given the freedom to explore resources such as Polydron to create their own shapes; naming such creations can be somewhat more problematic, however! Rather than worry about this, encourage the children to draw on what they know about the properties, such as number and type of faces, and look for similarities and differences with known shapes.

Shape orientation

As well as type of shape, the orientation in which they are presented can also play a part in reinforcing or dispelling stereotypical views. The equilateral triangle discussed earlier is typically drawn with a horizontal base, and, if this persists, children come to expect this of triangles. Ability to recognise those which do not have horizontal bases as triangles (and not 'upside-down' triangles, just triangles) is crucial to having a sound understanding of shape. In research by a past student (Knee, 2010) all 25 children recognised a square presented with a horizontal base, but this fell to only three of the children when the square was presented on a tilt as we see in Figure 10.6.

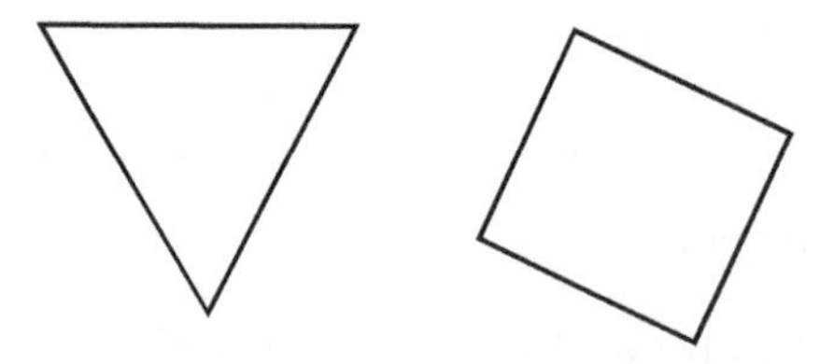

Figure 10.6 Atypical shape orientations

Pyramids are a slight exception to the orientation rule in that they are described in relation to their base, for example, a pentagonal-based pyramid with five triangles coming up from the edges of the pentagon and meeting at a point. The shape does not stop being a pyramid when orientated differently, however; it is just conventional to describe it in this way. Picturing real-life examples of cones may help to demonstrate this point; I visualise cones in different orientations depending on whether I have in mind a traffic cone or an ice cream cone!

One last point about orientation relates back to the use of junk boxes. Whereas a mathematical cuboid has no top or bottom, front or back, my cereal box does. There is therefore some mileage in discussion of these sorts of differences and the comparison of mathematical and everyday language with children.

The language of shape

Mathematical terms to avoid ambiguity

Use of specific mathematical language to describe the properties of 2D and 3D shapes helps to avoid ambiguity. The term which causes the most confusion regarding shape is 'sides'. Number of sides seems entirely appropriate when talking about flat shapes, particularly with young children, although even here the everyday use of 'sides' may involve turning a shape over to see the other side, and thoughts about things like front and back. With solid shapes, 'sides' is typically avoided, being replaced by 'faces' to describe the flat shapes (or surfaces) the solid is comprised of, and 'edges' to mean the lines along which the faces join. 'Corners' gradually become 'vertices' (the plural of 'vertex'), which is helpful at a later stage when children may well consider the angles at the corners of the 2D faces in relation to the vertices of a 3D shape. But in primary classrooms I find that children use corners and vertices interchangeably without it causing too much of an issue.

In an earlier section we met the term 'polygon' to describe 2D shapes with only straight sides and met some of the solid shapes, such as the tetrahedron or triangular-based pyramid. The generic term for 3D shapes, often described as 'solids', is 'polyhedron' for a single solid shape, and polyhedrons (or polyhedra) in the plural. Note that just as a polygon has only straight sides, a polyhedron only has flat faces consisting of polygons.

Terms like 'cube' and 'cuboid' are clearly linked linguistically, and here we face a similar issue to that which we explored in relation to squares: a cube is, by definition, a member of the cuboid family since it has six rectangular faces which just happen to be squares. Any cuboid (your typical box shape), and therefore any cube, is also a rectangular prism: whichever way you look at it, the end shapes are parallel and the same and they are joined by rectangles.

Congruence and similarity

Congruence is a term you may have heard in relation to 2D shape. When shapes are said to be congruent, this means that they are mathematically the same. This may be obvious, or it may be necessary to turn one of the shapes round or over to check that it really is the same, and if it is, then one shape will fit exactly over the other. Rather than being congruent, 'similar' shapes are those which are bigger or smaller versions of the same shape, having been scaled up or down in size. Note that the angles remain the same in this instance, with just the sides changing in length. For more on this topic, see Haylock (2010).

Tessellation and tiling

Tessellation and tiling are aesthetically pleasing aspects of shape work but there may at times be a danger of the mathematics getting lost, so let's just think about what that mathematics would be. Whilst sources seem to vary, I usually work from the premise that 'tessellation' is where the *same* shape can be repeated over and over again to cover a flat surface, leaving no gaps. The only regular polygons it is possible to do this with are triangles, squares and hexagons as we see in Figure 10.7. The mathematical reason behind this is down to the angles involved of course: six triangles with interior angles of 60° can fit around a point (360°) leaving no gaps, as can four squares ($90° \times 4 = 360°$) and three hexagons ($120° \times 3 = 360°$). If we try the same task with pentagons, for example, we find that three pentagons are not enough (we are left with a small gap because $108° \times 3$ is less than 360°) but four would not fit (because $108° \times 4$ is greater than 360°).

Figure 10.7 Tessellation around a point

Once we try irregular shapes, however, we find that we can tessellate some pentagons, and in fact we discover that any triangle or quadrilateral will always tessellate. Both of the following sites: http://library.thinkquest.org/16661/simple.of.non-regular.polygons/quadrilaterals.html and http://www.mrbutler.org.uk/world/demo/tessellate-quads.html demonstrate the fact that quadrilaterals will always tessellate and will also help you to understand how tessellation works.

Using more than one shape to cover a plane is generally referred to as 'tiling' to distinguish it from tessellation involving a single shape. Patterns involving octagons are a good example of tiling as they can be combined with squares to make an attractive floor covering or similar.

At a later stage in their mathematics education, children may begin to realise that there is a connection between tessellation/tiling, where the corners of 2D shapes join around a point on a flat plane, and the vertices of 3D shapes where the 2D faces come together. Try picking any vertex of any solid shape: what do you notice about the angles involved? (See Self-Test Question 2 for further information.)

Activities for shape teaching

Whilst 2D shape lends itself to use of the interactive whiteboard, when teaching about 3D shape ensure that children are given opportunities to handle actual solids; pictures of solids are generally a poor substitute!

Some of the following activities target just 2D or 3D shape, whereas others mix the two. For the most part the activities are very simple ones, easy to differentiate and to adapt for children of different ages, for example by focusing on alternative shapes. When planning to use these activities, consider which part of your lessons they will prove most useful in. Aim, as always, for quality time spent focusing on particular aspects of shape rather than trying to get more done but less well. Discussion will ideally play a big part.

Pupil activity 10.1

Dotty shapes (2D)
Y1 to Y6

Learning objective: To recognise different versions of 2D shapes in different orientations.

The following activity can be adapted to any 2D shape and also any age group. The illustration here relates to triangles but is just an example, and you might choose to explore different types of angle, and so on.

Mark 15 or so dots on the board and ask the children to imagine joining some of the dots to make a particular type of shape specified by you. The older the children, the more you can ask them to visualise their shapes; with younger age groups, progress more quickly to actual drawing of the shapes. Ask the children to spot different versions of the same basic shape, planning in advance the sort of language you want to promote. For example, with the youngest children you might just want to reinforce the fact that they have made a triangle as long as it has three straight sides ('let's count and check') and that you can have big triangles, or skinny triangles, or whatever, and that orientation does not matter. With the older children you might want to revise different triangle types ('does anyone think they can spot an isosceles triangle?') whilst at the same time reinforcing the fact that orientation does not matter. Children interacting with the dots and discussion of the shapes generated are both key to the learning here, and this activity can lead nicely into some work on paper, either with random dots or switching to exploration of more typical grids (see activity 10.2).

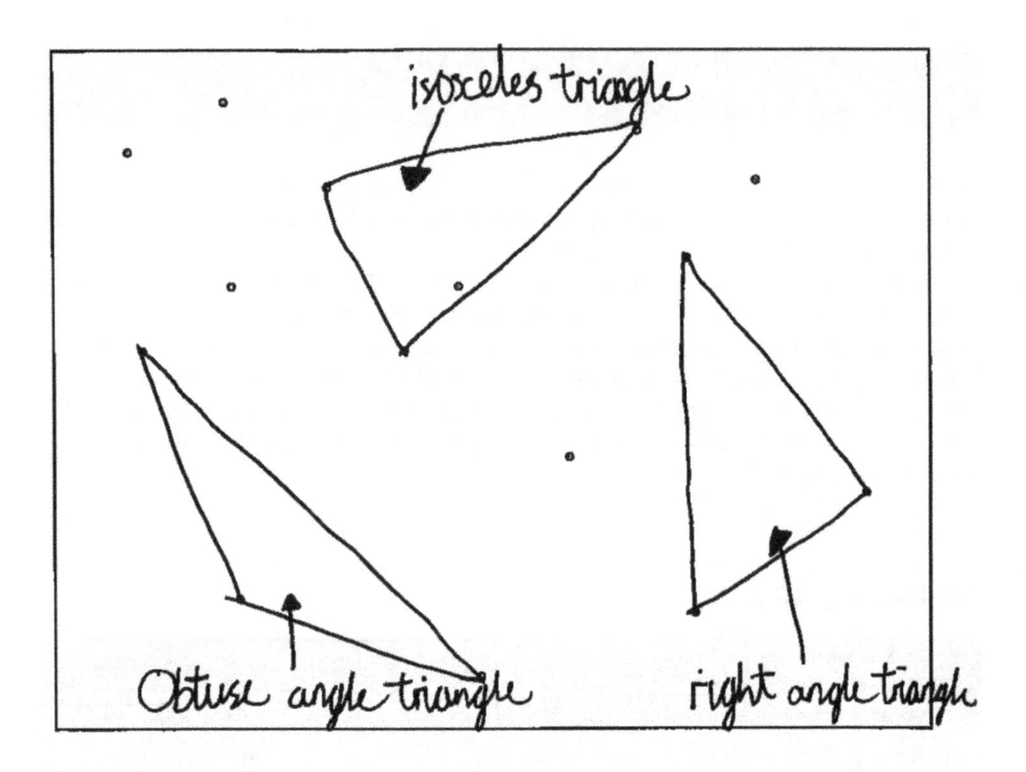

Pupil activity 10.2

How many shapes can I find? (2D)
Y1 to Y6

Learning objective: To appreciate the standard properties of a given shape whilst recognising alternative versions.

Regular grids, either dots on paper or geoboards (pin boards used with elastic bands), are also a useful resource, particularly for exploring different examples of the same shape. Here we see a variety of parallelograms (noting the presence of rectangles), but the focus could be on other four-sided shapes, or different sorts of triangle, or terms like 'perpendicular', for example.

If you want children to answer questions about how many different types of parallelogram are possible, it is best to limit the grid, for example, using just nine pins in a three-by-three arrangement; it helps to be able to record the shapes discovered. This may also be a good time to introduce the idea of congruence, checking that you have not mistakenly counted the same shape more than once just because you've drawn it a different way round!

Note that we get isometric grids with pins in a triangular arrangement as well as pins arranged in squares as we see here. Which type of geoboard would you want to use to make an equilateral triangle and why? (See Self-Test Question 3.)

Pupil activity 10.3

And another (2D)
Y1 to Y6

Learning objective: Ability to recognise essential as well as transient properties of shapes.

This is a very simple activity detailed in various sources (see, for example, ATM on YouTube, via http://www.atm.org.uk/community/) and is based on the idea of children drawing a particular shape (2D) and then being asked to draw another but to make it different from the last one. Comparing the children's pictures allows discussion of what constitutes the shape under consideration and to potentially find atypical examples, such as a quadrilateral with a reflex angle. As with most activities it can be adapted to suit a particular purpose, such as introducing the children to the idea of mathematical similarity.

Opportunities can also be taken to present the children with decoy examples in order to debate whether the essential properties are illustrated.

Pupil activity 10.4

Elastic shapes (2D)
Y1 to Y6

Learning objective: Ability to use the language of shape to articulate 2D shapes made.

What sorts of shapes can you make with an elastic band? What if you had a friend to help you, an extra pair of hands? Such a simple activity, but one which can give children opportunity to explore different versions of given shapes and to rehearse with their partner the language associated with 2D shape.

A large piece of elastic helps to move the discussion from the pairs to a larger group, and you can direct the questions to focus on whatever aspects of shape you have in mind for your class. Elastic works well because of the 'give', but you can use string instead if that is what you have available.

Pupil activity 10.5

Geo Strip shapes (2D)
Y1 to Y6

Learning objective: Ability to make and describe 2D shapes.

Geo Strips are a plastic resource, not unlike Meccano, and pieces are joined using split pins. They are ideal for making and modelling different examples of 2D shapes, for example, mathematically similar triangles. With the younger age groups I tend to use them predominantly to show and pass round shapes I have made, whereas the older children can use them to investigate particular shapes for themselves. Selecting only particular lengths for use or stipulating certain rules helps to focus the learning. For example:

- Make a square... can you change it into any other shapes without taking it apart?
- Make a pentagon... now can you make it look different using the same strips?
- Which combinations of three strips allow you to make right-angled triangles? How many different examples can you find?
- Are there any three strips you could choose which would not make a triangle?!

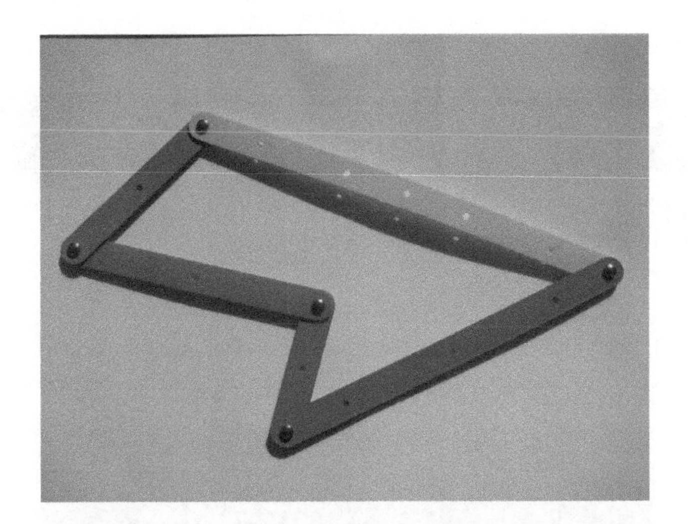

Pupil activity 10.6

See-through shapes (3D)
Y2 to Y6

Learning objective: To develop appreciation of the structure of solids with reference to the shape's edges.

With Polydron (or a resource called Clixi) the emphasis is on the faces of the 3D shape; with straws, using pipe cleaners or modelling clay to join the corners, the focus switches to the edges of the faces and solid. The added advantage here is that you can see through the shapes very easily to get a better sense of their structure. Their use can focus on any 3D shape of your choice, and 2D shapes can also be made of course.

Pupil activity 10.7

Shape investigation (3D)
Y3 to Y6

Learning objective: To appreciate that different solids can have the same number of faces without being the same shape.

If a 2D shape has three straight sides, then it is a triangle, but with solids we find a little more variety. Using Polydron or similar, set the class the challenge of making as many different shapes as they can think using exactly five pieces as faces, progressing to six, and so on.

Once the shapes are made, an interesting side line is to compare the number of faces with numbers of edges and vertices. Here we have two shapes with five faces, a triangular prism (6 vertices and 9 edges) and a square based pyramid (5 vertices and 8 edges). As we can see, these numbers change depending on the arrangement of the faces, though later the children will hopefully discover, as Euler did, that there is a relationship!

Pupil activity 10.8

Shape shopping lists
Y2 to Y6

Learning objective: Ability to relate 3D shapes to their constituent 2D faces.

Polydron and Clixi offer wonderful opportunities for children to explore the relationship between 2D and 3D shapes, and time just to play and build is of course beneficial. But at other times you will want to find ways of focusing their concentration on particular aspects such as the individual faces. One nice way of doing this is to have the children envisage the solid shape they want to make and to list the individual pieces they want to 'buy'. For example, if I ask for four triangles and a square, I will be able to make a square-based pyramid. Older children, using more advanced sets of Polydron, can begin to specify certain types of triangle for what they have in mind. Understanding grows when you get back to your seat to find you are missing a piece for the solid you want to build! Differentiation occurs quite naturally in the choices children make about the shapes they want to build, from quite simple to incredibly intricate solids. The activity is also good for purely practical reasons: not everyone needs the equipment at the same time!

Pupil activity 10.9

2D to 3D through nets
Y3 to Y6

Learning objective: To develop understanding of the relationship between a solid shape and its faces.

Having explored the properties of the junk boxes and become familiar with building solid shapes using construction equipment, the next step for many children will be to create their own nets using cardboard. I strongly recommend children attempting this from scratch rather than being supplied with a template; depth of learning often mirrors the level of challenge involved in the activity. Choice with regard to what to make is also important since this allows the child to self-differentiate and allows those with good spatial ability (not always your typical high fliers) to really shine. If anyone is struggling to visualise the required net, either building a replica from something like Polydron and taking that apart, or deconstructing a junk box, should help the child to get a better sense of the layout of the individual faces.

Pupil activity 10.10

Shape sketches (3D)
Y3 to Y6

Learning objective: To appreciate that pictures of 3D shapes are actually 2D.

Advances in digital photography make it so easy these days to represent solid shapes in two dimensions by taking a picture of them, but having the children attempt to capture the qualities of a solid shape in a sketch helps them to appreciate the structure and properties of the shapes to a greater extent. Working in small groups, children can sketch from different angles and be prompted to discuss which sketch seems to capture the properties of the shape the best, stressing that looking at the picture should evoke the 3D nature of the object.

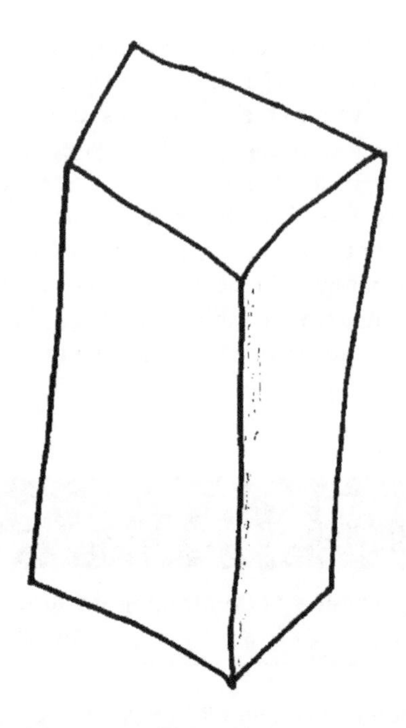

As the children grow older they will begin to appreciate that solids are presented in particular ways on paper as this helps to convey all their features more clearly.

Pupil activity 10.11

3D shape similarities and differences
Y1 to Y6

Learning objective: Ability to recognise the properties of a solid shape that define it and make it similar to or different from other 3D shapes.

Collect or ask the children to bring in a variety of junk boxes; check that you have the examples to illustrate any particular points you wish to ensure are raised, which of course will vary according to the age group of the children. Consider whether 'special' examples have been represented (cubes as members of the cuboid family, cuboids as members of the prism family, etc.) in order to reinforce rather than overlook some of the family similarities. You may also want to include some 'decoy' shapes such as an 'almost' cube.

Give each child or pair of children a box, and everyone then circulates the room comparing their box with those of their classmates. What properties do they notice that their shapes have in common and what are the distinct differences? Whilst you will want to model the accurate use of any terminology at various points, the style of activity presented here helps to give children ownership of the vocabulary, and discussion may well involve, for example:

- curved and flat faces;
- numbers of faces, vertices and edges;
- names of shapes and shape families;
- colour, size and contents of box.

This activity is a good precursor to sorting the shapes and discovering that many of them belong to more than one shape family, an activity I like to do with the whole class seated in a circle, but you may feel it would work better differently with your class. Here I ask pairs of children to select a shape and talk to each other about what they can say about it – this gives them time to prepare their contribution. Then as they show and talk about their box, guide the class in comparing and contrasting the shapes, support them in modelling suitable vocabulary (faces, edges, corners/vertices, parallel, perpendicular, etc.), praising their attempts to use new vocabulary and inviting comments from other children if appropriate. The activity is inclusive since every child can say something about their box, even if it is just the colour and even if they rely on everyday language, whilst differentiation is relatively straightforward if a range of more complex shapes is available. Do not worry if children have to revert to saying 'It's a bit like . . . ' when they do not know the name of something. Believe me, this happens in my lessons too; the weirdest shapes provoke quality reasoning and the most wonderful discussions!

Pupil activity 10.12

Sorting shapes into families
Y1 to Y6

Learning objective: To recognise that shapes can be members of more than one family.

Children need lots of opportunities to sort shapes according to their properties, and with some careful planning can be forced to recognise overlaps between shape families by limiting the categories on offer. If I have a selection of polygons including squares, and a choice of sets labelled 'triangles' and 'rectangles', then I have to decide whether my square fits into one of these sets or not. Collaborative working is key to the success of this type of activity as children force each other to re-evaluate their decisions through discussion, particularly if they end up arguing about where a shape belongs! Carroll diagrams also lend themselves to sorting shapes into families and, like Venn diagrams, can focus on 2D or 3D. Where would familiar shapes end up in the Carroll diagram example here?

	rectangular	*not* rectangular
all same length sides		
not all same length sides		

An alternative approach is to present the children with a diagram which has already been completed, for example, intersecting sets of a Venn diagram labelled 'squares' and 'rectangles'. This should result in an interesting discussion, since the purely 'squares' section would be empty!

Pupil activity 10.13

Chains of reasoning based on a shape's properties
Y2 to Y6

Learning objective: To appreciate the role a shape's properties play in defining what it is.

Sorting trees can be useful in identifying different types of shape, but these need to be chosen or created with great care, as I have seen examples which lead to incorrect answers. In many ways there is more mileage in children working together to produce their own, checking carefully that their reasoning works for any shape that anyone can think of.

Pupil activity 10.14

Shape true or false
Y1 to Y6

Learning objective: Ability to apply shape knowledge to correctly identify true and false shape facts.

This activity requires shape statements to be prepared in advance. The children should work collaboratively to decide whether they agree or disagree with them; this is a great way of getting them to articulate their understanding about shape, potentially modifying their understanding as a result of discussion. Statements can be contentious to deliberately provoke discussion – how would you respond to the final statement below? The style of activity recognises explicitly the presence of misconceptions. Some suggestions are given below and range from basic identification for young children to more complex issues for older children. You will no doubt be able to add to the list.

- This is a triangle True or false?
- A seven-sided shape is called a septagon. True or false?
- A pyramid always has five faces. True or false?
- A parallelogram can never look like a square. True or false?
- A triangle can never have an obtuse angle. True or false?
- A sphere is the shape of a ball. True or false?
- A circle is a polygon with an infinite number of sides. True or false?

Note that concept cartoons would be an equally good way of exploring misconceptions relating to shape; but whatever your approach, your ability to take a role in mediating the discussion is crucial.

Pupil activity 10.15

The concept of 'circle'
Y3 to Y5

Learning objective: To develop a deeper understanding of the definition of a circle.

By definition, the outline of a circle is a set distance from the centre of that circle, and a simple but effective way of exploring this is to play a beanbag game with the class or a large group. (It works less well if the group consists of only a few children.) Line everybody up in the playground or hall and let them try scoring a point by throwing a beanbag into a bucket placed somewhere in front of the line. What you hope will happen is that someone will start moaning about the fact that it's not fair, that you have more chance of getting the beanbag in (and gaining a point) from certain positions in the line than others. Let this progress into a discussion about what the fairest arrangement would be, and of course being equidistant from the bucket would require everyone to be standing in a circle!

Pupil activity 10.16

Paper folding shapes (2D)
Y1 to Y6

Learning objective: To develop the confidence to describe known and unknown 2D shapes.

Each participant or pair in the activity will need their own supply of pieces of paper, with regular A4 (or A5) ideal. There are two different ways in which this activity can be approached: either the children make a fold and then open the sheet out again to see the shapes created on either side of the fold (perhaps inking in the fold line to make it more obvious), or the paper remains folded and the outline of the paper becomes the focus shape as pictured here. The original idea came from Barmby *et al.* (2009: 137).

Key to the success of the activity is the quality of the teacher's questioning, and suggestions include:

- How many sides has it got? Has anyone got one with more sides?
- What sort of shape have you made? What other polygons are possible?
- What are your angles like? Any particularly huge or tiny ones?
- Does anybody's shape have any lines of symmetry?
- What range of quadrilaterals could be made? What about pentagons, hexagons and so on?

Like many of the other activities, lengths of sides, sizes of angles, and so on could then be measured if that suited your purpose.

Pupil activity 10.17

Find and fill (2D)
Y1 to Y6

Learning objective: Ability to identify irregular versions of given shapes.

This activity arose to counteract the prototypical shapes children are often presented with. The starting pattern can be created in a variety of ways, such as using an ICT art package (here I used something called 'Dazzle'). You will probably want to make sure that certain shapes are well represented in your random pattern, depending on the focus of your lesson!

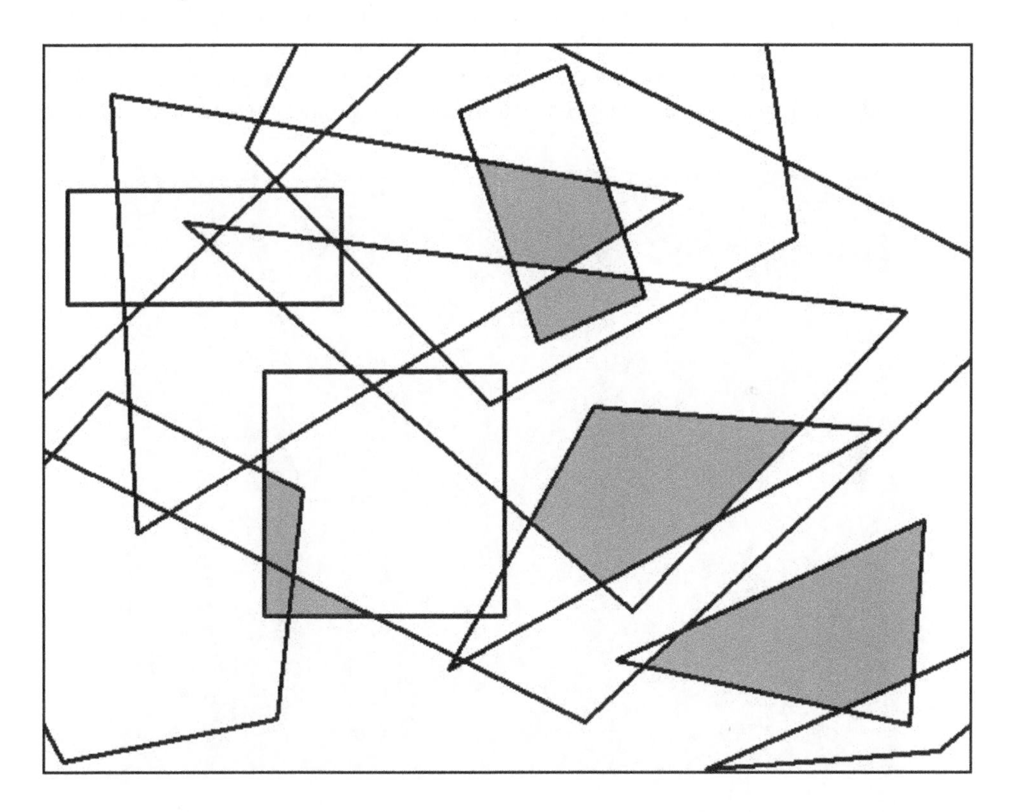

This is best introduced as a whole-class or group activity as the emphasis is on identifying and talking about different families of shapes together; as you can see above, some pentagons have been shaded, taking care not to shade any 'almost' pentagons! The one on the left is a combination of shaded sections which made a pentagon overall. The same basic activity could be adapted to create independent work, but be wary of losing the opportunities for discussion.

Pupil activity 10.18

Tessellation and tiling
Y1 to Y6

Learning objective: Knowledge of which shapes fit together without gaps and developing understanding of why this works.

Commercially available pattern blocks are ideal for learners of any age exploring tessellation or tiling, but older children can also be given opportunities to generate their own shapes. The simplest way to ensure that any changes made will result in a shape which still tessellates is to start with a shape which tessellates and replicate any changes to one side on the opposite side as seen in the design below. This was created by making a cardboard tile and drawing around it. Once children have understood the basics they may well want to experiment with more complex ideas or investigate the works of M.C. Escher and his followers. The website www.tessellation.org may be useful in this respect.

Pupil activity 10.19

Close-up mathematics
Y1 to Y6

Learning objective: To develop attention to shape-related detail.

Close study of detailed photographs and posters, including those specifically designed to support mathematics teaching, is a good way of focusing children's attention on specific features of mathematics – in particular, aspects of shape. The example here was chosen because of its potential with regard to both 2D and 3D shape. This style of activity particularly helps to promote careful observation as well as discussion with peers – an opportunity to rehearse mathematical language. Spatial awareness is also a key feature of attempting to reproduce the picture shown. Children will need to be in small groups, each group having a large sheet of paper and a pen. Pen seems to work better than pencil as there is no temptation to rub anything out and because the resulting picture will be bolder. The most suitable group size will vary depending on the age of the children. The picture needs to be displayed out of general sight, with one individual from each group invited to view the picture for one minute before returning to their group to talk about what they have seen and to draw what they can remember of the picture. As each group member takes their turn, the detail of the picture should develop. Once everyone has had a turn, allow the children a little extra time to finish their picture, at which point the groups can compare their picture with those of the other groups. Once they see the original picture they should be more than ready to discuss the mathematical features you want to elicit. I am indebted to an in-service history course many years ago for the original idea. As well as being good for history and mathematics, the activity could equally well be used to draw children's attention to features of geography or science, for example.

Summary of key learning points:

- by providing rich ranges of 2D and 3D shapes, we offer learners the best chance of developing appropriate and inclusive concepts regarding different shapes;
- both regular and irregular shapes should be presented in different orientations;
- 'decoy' shapes (those 'almost but not quite' examples) need to be incorporated to help children to develop the ability to discriminate between the decoy shapes and the real thing;
- overlaps between shape families need to be overtly drawn to learners' attention.

Self-test

Question 1

The prefix 'semi' implies half of a shape, thus we get 'semicircles' and 'semispheres'. True or false?

Question 2

For shapes to be able to tessellate (or tile) across a flat surface the angles at each point must add up to 360°. However, for the vertex of a solid shape, the angles involved must add up to less than 360°. True or false?

Question 3

If I want to make an equilateral triangle on a geoboard using an elastic band, I could only use a board with the pins in an isometric arrangement. True or false?

Question 4

If I want to make a triangle out of Geo Strips (or similar), I can do so whichever three strips I pick out of the box. True or false?

Self-test answers

Q1: (False) Only one of the terms is correct. We have 'semicircles', but half a sphere is referred to as a 'hemisphere'.

Q2: (True) Once the angles around a point add up to 360° we know that there will be no gap and the shapes will lie flat; to create the vertex of a solid shape we need at least a small gap as it is the joining of the edges across the gap which creates the vertex or point. You might want to play around with this idea using Polydron; for particularly 'pointy' vertices, try three equilateral triangles joining at a point (you will have made part of a tetrahedron). What happens to the vertex if you use more than three triangles?

Q3: (True) With the pins in a square arrangement I could make a shape that looked a bit like an equilateral triangle but if I checked the measurements (lines or angles), I would find it was not quite equilateral.

Q4: (False) I will probably be lucky and pick out three strips which will make a triangle, but this depends on the lengths I choose. If the two shorter lengths combined are not longer than the longest strip, then I will not be able to fix my triangle together. It would be like trying to draw a triangle with lengths of 3 cm, 4 cm and 8 cm: impossible!

Misconceptions

'All triangles look the same'

'A square is not a rectangle'

'There are lots of regular quadrilaterals'

'Right angles always look like an "L" shape'

11

Space

About this chapter

This chapter describes the three main topics within 'space', the spatial element of the mathematics curriculum. We make links back to Chapter 10 on shape where appropriate and explore symmetry. The chapter is divided into the following sections:

- Position and direction
  - Positional and directional language
  - Compass points and directions
  - Coordinates
- Movement
  - Transformation
  - Enlargement
- Symmetry

What the teacher and children need to know and understand

Spatial learning in mathematics relates to three main topics:

- Position
- Direction
- Movement

Spatial ability generally is believed to support such skills as being able to read or tell the time (Ryan and Williams, 2007).

Position and direction

Positional and directional language

Children's earliest experiences of position stem from the use of positional language in everyday activity and through play; for some children this will be well developed through interaction at home. This terminology can be gradually refined through hearing words modelled by adults and through being given opportunities to use the language themselves. The language includes terms such as 'above', 'below', 'on', 'in', 'under', 'underneath', 'on top of', 'next to', 'in between', 'in front', 'behind' – a list of potential vocabulary which is already pretty long but which I am sure you will find you can add to!

As with position, the focus of direction work varies considerably according to the age and stage of development of the learner; use of 'left' and 'right' comes to mind as something even some adults find tricky. Eventually one's familiarity with left and right should progress to an appreciation that defining left and right can be particularly awkward when people are opposite each other. This has implications for modelling left and right, ideally facing the same way as the children.

Note that we actually use left and right in two subtly different ways; I can be talking about turning to my left or right, or I might be referring to the left-hand or right-hand side of something. You will need to be clear which is your intended focus in order to analyse carefully any teaching materials you find, ensuring they will have the correct emphasis. I have noticed that this is particularly problematic with interactive whiteboard materials such as one I saw recently where a mouse was navigating an on-screen maze to get to some cheese. Navigating a maze in reality involves turning to the left or right, deciding to go straight on or to back track. The software instead required the user to focus on the mouse getting to the left- and right-hand sides of the board to reach the cheese, thus failing to mimic a real journey through a maze.

In the story above the software had been designed so that the mouse did at least turn and face the way it was travelling, reflecting the fact that we would tend to look the way we are going in a maze situation. Difficulties arise if the picture of the moving character is fixed and always facing in the same direction. Picture a pirate navigating paths on a treasure map on an interactive whiteboard, supposedly turning left and right to take different routes. However, because the pirate picture is fixed, trying to determine whether the pirate should turn left or right becomes a considerable issue when he is facing backwards at a junction! Even if the pirate did turn as the mouse did, you would still notice children craning their necks and struggling to work out whether the character wants left or right when the maps and mazes are presented in a fixed position on a vertical board. Working horizontally is always better for work such as this, and many children and adults benefit from being able to turn either themselves or the map round.

Story books and scenarios involving journeys lend themselves to this topic; describing the route the postman takes on their delivery round, for example. But given the issues raised above, take every opportunity to act out the scenario with the children or to draw routes out horizontally rather than relying on the interactive whiteboard.

In the very early stages directional concepts relate very much to the viewpoint of the child who is beginning to appreciate that things are in front of them, behind them, and so on. This relates to movement forwards and backwards and other directions travel can take. Eventually specific terms in addition to left and right such as 'clockwise' and 'anti-clockwise' help to define direction of turn using convention related to analogue clocks.

Compass points and directions

Work involving navigational compasses raises similar issues to those associated with left and right and is problematic when taught on an interactive whiteboard; since when has north been 'up'? Difficult for left and right, the vertical orientation of the board is also quite inappropriate for compass work and should generally be avoided, certainly to start with. Work with the children at ground level and orientate north, south, east and west to roughly the directions they actually lie in, perhaps talking about the north and south poles in conjunction with looking at a globe. East and west are more conceptually challenging but fortunately, where children are concentrating on compass directions at a local level, appreciating these as fixed directions tends to be sufficient. Once north, south, east and west are 'fixed', the points between are generated using combinations of them, as shown in Figure 11.1.

Whilst most work will focus on direction of travel towards whichever compass point, you should be aware of a subtlety in regard to compass directions. Picture yourself walking east early one morning and enjoying a beautiful sunrise. Now imagine another walk with a stiff easterly breeze. Whilst both refer to the east, you are walking *towards* the east whereas the wind is coming *from* the east. Since real life is full of these little complications, endeavour to ensure that your mathematics teaching takes them into account!

Common misconception

'As the wind is going south it is called a southerly wind'

Winds are referred to according to where they originate rather than where they are going, therefore this child is in fact describing a northerly wind.

Figure 11.1 Eight points of the compass

Coordinates

At a much later stage, position becomes linked to coordinates: the idea that I can specify where something is using mathematical conventions. Coordinates are initially the type we might expect to find on a street map, with the horizontal axis labelled using a letter and the vertical axis a number. This results in being able to identify a particular square; in B2 in Figure 11.2 we find a public house and various sporting opportunities! When modelling this with children I tend to number from the bottom upwards as this matches the direction of later coordinates, but I do not think it really matters.

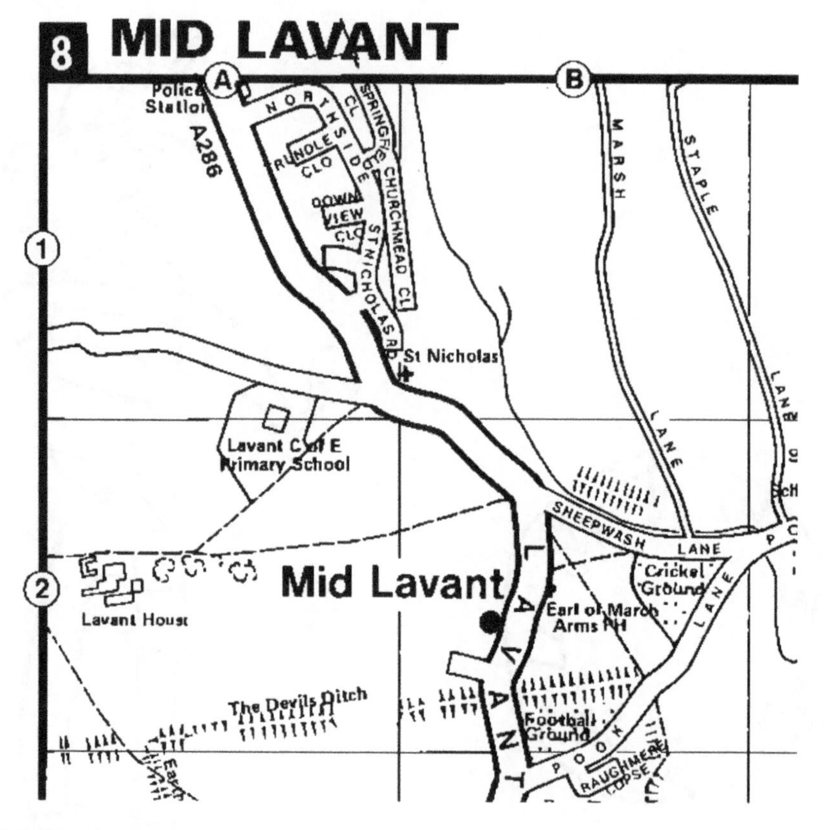

Figure 11.2 Street map
Reproduced by permission of Ordnance Survey on behalf of HMSO.

Cartesian coordinates (named after their inventor, René Descartes) differ in that they allow us to identify a point using a pair of coordinates, with the x coordinate (for the horizontal axis) given before the y coordinate (the value to be read from the vertical axis). The x and y values are conventionally presented in brackets and separated by a comma, for example (3, 5), with (0, 0) referred to as the 'origin'. With practice it will become ingrained which way round the coordinates should be read, but many children use 'along the corridor, up the stairs' type sayings to help them remember which way round they are. Clearly, if you want to check that children are getting the hang of this, then you will want to avoid examples like (5, 5). Coordinate activities regularly involve plotting points and joining them up, typically working in only the first quadrant (using positive numbers) with primary-age children.

Common misconception

'I can plot all my coordinates and then join them up in whatever order I like as it won't make any difference'

What children will find out is that actually the order does matter! I am safe with triangles, but with four or more points I have to know which order they are supposed to be joined in. This was supposed to be a square!

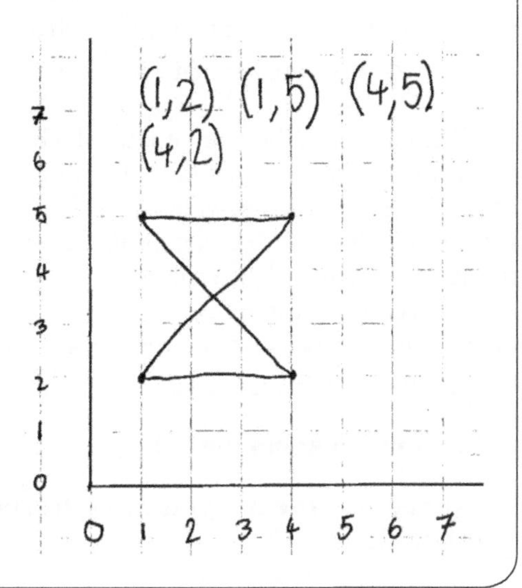

If we were to use a map analogy here, this style of coordinate is more akin to what we would find on an Ordnance Survey map; however, grid references are not given in the same style as Cartesian coordinates. Each map has a sheet number, and position on the map is determined by a six-figure grid reference. For example, the other day I walked from 071125 which involved finding the line 07 on the horizontal axis and going one-tenth into the box (the 071 part) and then locating the line 12 on the vertical axis and going about half-way into the box (the 5 of the 125, equivalent to five-tenths). The car park was just where the map suggested it would be!

Clearly there are strong links between mathematics and geography. These days, children's ability to appreciate the bird's-eye view that maps depict is enhanced by access to things like Google Earth; enjoy using the tools available to you!

Movement

Transformation

Mathematics can help us to describe the movement of things, with the term 'transformation' used to describe the various types of movement. Here I am imagining that it is a shape being moved. Note the links with congruence; whilst the new image may be in a new position, it is congruent to the original image. The three main types of movement are:

- rotation (a turning movement resulting in the shape being in various different orientations);
- translation (a sliding movement where the orientation of the shape is retained);
- reflection (where the shape flips over and may therefore appear as a mirror image).

With rotation, the point about which the shape is to be turned has to be specified and varies; for younger children I think it is easier to grasp if it is in the middle of the shape, but it can also be at the corner of the shape, or completely away from the shape. By the time this is the case, coordinates are typically used to describe the starting and finishing positions of the shape.

Reflection is further explored in the symmetry section below, and for a more in-depth look at transformation topics, see Rowland *et al.* (2009).

Common misconception

'Reflection means you have to repeat one side of the page on the other side'

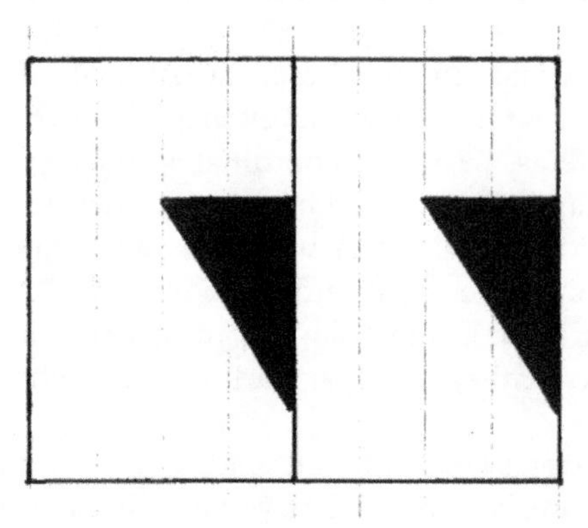

Whilst reflection does require a repetition of sorts, it needs to be a mirror image, maintaining an equal distance from the line of symmetry.

It would be wrong to give the impression that movement is only about the transformations described above. From the perspective of a young child, movement is about getting to grips with yet more language: forwards, backwards, sideways, up, down, and so on. There are also links with some of the measurement topics discussed in Chapter 7, since movement can also be described in terms of time and distance: 'awful traffic jam this morning; took me about 20 minutes to do a mile'.

Early rotational movement is defined in terms of how far you turn, with whole turns taking you all the way round, half-turns leaving you facing the opposite direction and so on. Like many of the 'space' topics, this is best taught dynamically; get the children up and moving around. Children progress eventually to defining the size of the turn in terms of degrees rather than fractions, affording greater accuracy.

Enlargement

Typically included alongside transformation is that fact that shapes and other images or items can be made larger and smaller, with enlargements by numbers greater than 1 resulting in the shape getting bigger, and enlargements by values less than 1 resulting in the shape getting smaller (noting the slightly counter-intuitive use of the term 'enlargement'). For example, if a square is enlarged by a scale factor of $\frac{1}{2}$, each of its sides will be halved in length.

Common misconception

'As the lines are twice as long the area must be twice the size too'

Enlarging something by a scale factor of 2 means that the lengths involved will double. In the example here the dotted lines may help to illustrate the resulting area change, with the enlarged triangle *four* times bigger than the original (2^2 bigger). An enlargement by a scale factor of 5 would result in an area increase of 5^2; squared because area is two-dimensional.

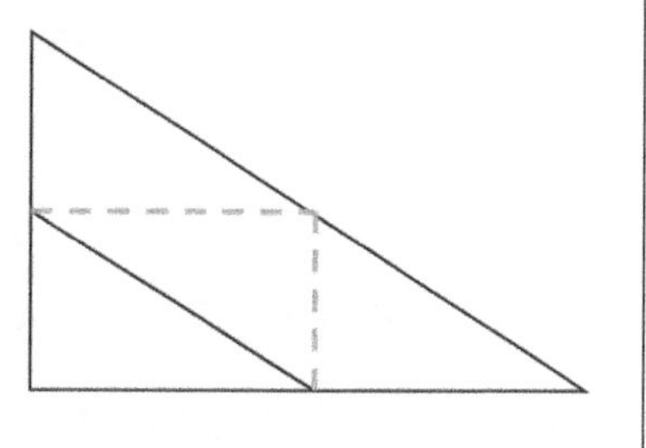

Symmetry

It was a dilemma which chapter to include symmetry in; it is a topic which often draws on shape, but could be considered to be about position and movement since appreciation of symmetry relates to visual and spatial ability. Different types of symmetry exist:

- Something is said to have rotational symmetry if, as it rotates about a point, it looks exactly the same in other positions and before it gets back to the start. By this definition something has no rotational symmetry if it has to go all the way round before it looks the same again.
- Something is said to have reflective symmetry if a line can be drawn such that one half of the object or pattern is a mirror image of the other.

Rotational symmetry is described in terms of 'order of rotation', and here we find some differences of opinion on how to count this. If something looks the same in three positions, then it is said to have order of rotation 3; but if something has no rotational symmetry, some mathematicians refer to this as being order 0, whereas others go for 1. See activity 11.17, where rotational symmetry is explored through photographing hub caps.

Reflective symmetry can involve mirror images of complete shapes, pictures or patterns (see activity 11.16), but we can also look for lines of symmetry within images and shapes, and it should be noted that shapes and patterns can have more than one line of symmetry (Figure 11.3).

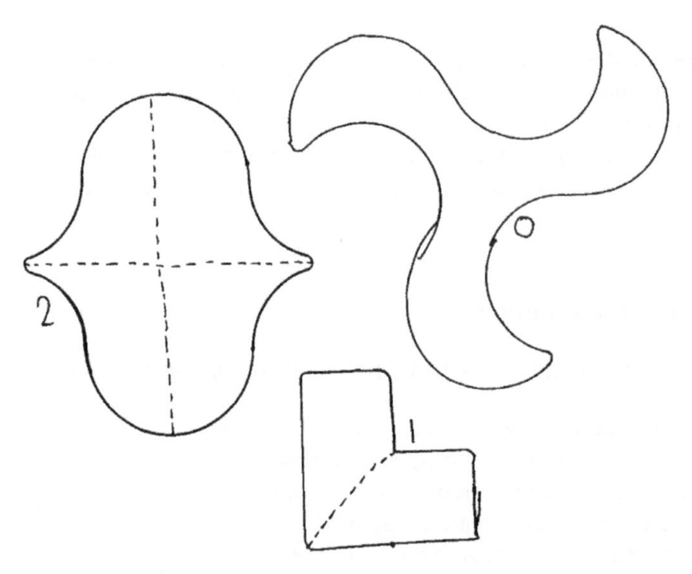

Figure 11.3 Different numbers of lines of symmetry

In early work on symmetry children will find it easier to spot symmetry or lack of it where the lines are presented horizontally or vertically, but need to gain experience of lines of symmetry in other orientations. Note that in 3D we would be talking about 'planes' rather than 'lines' of symmetry; this is potentially a little confusing where real-world images are annotated to show lines of symmetry. A picture of a house may have a line of symmetry but a real house does not!

Common misconception

'I have cut this shape in half so it must be symmetrical'

The relationship between halving and symmetry is an interesting one. If I find that a shape has a line of symmetry, then the two parts created must be equal in size and are indeed half the shape. Children need to recognise, however, that a line halving a shape will not necessarily be a line of symmetry, as in the example here.

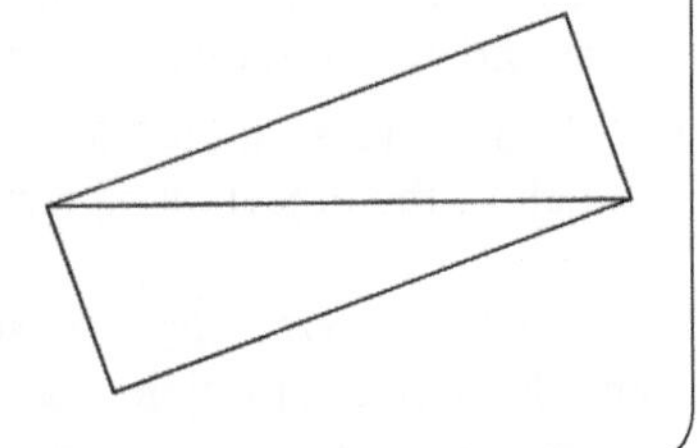

Activities for spatial teaching

When planning to use any of these activities, consider which part of your lessons they will prove most useful in, and whether practical activity would be more appropriate than interactive whiteboard activity. This may mean arranging to go out and about or to the hall. Remember that most of them can be easily adapted to meet slightly different objectives for different age groups.

Many of the activities in this chapter can also be successfully interchanged; for example, using the treasure maps in activity 11.5 for exploring compass directions instead of coordinates.

Pupil activity 11.1

Where is teddy?
Y1 to Y2

Learning objective: Ability to use and understand positional language.

Games played with a soft toy are one of the easiest ways to rehearse positional language, with teddy able to sit on things, under things, behind things, and so on. Try not to be tempted to do this type of activity on an interactive whiteboard where there can be so much more ambiguity as to how the terms might be interpreted. With the children, list as many different position words as they can think of; these can then be selected at random and acted out.

With slightly older children who are still struggling with positional language, models such as doll's houses or elaborate space ships offer similar opportunities to rehearse the language together. Photographs of, or trips to, something like a play park can also provide a useful starting point for discussion of terms associated with the language of position.

Pupil activity 11.2

Position, direction and movement in gymnastics
Y1 to Y3

Learning objective: To develop the language to describe position, direction and movement in PE.

Gymnastics and other PE lessons provide a particularly good opportunity for children to rehearse spatial language. Once you have in mind the terms you want to reinforce with the children, plan how you will incorporate them into your gymnastics lesson. For example, you might ask the children to plan a sequence with a specific element and to explain to a partner in advance what they are going to do (in order to give children the opportunity to use the language themselves). Movements to be incorporated might include:

- jumping over;
- swinging underneath;
- weaving in and out;
- travelling in between.

Pupil activity 11.3

Story book journeys
Y1 to Y4

Learning objective: To develop understanding of the relationship between maps and routes around them.

So many different story books lend themselves to map making, and once the map has been created, it can be used to illustrate journeys around it. You will no doubt already have some books and stories you might like to use; you will want to think about the features of position and direction to which you would want to draw the children's attention. Below I suggest some of my favourite authors/books, noting that the maps and journeys are not necessarily explicit in the stories but that the story provides the starting point. This means that the focus can be tailored to the age group and is easily adapted to meet different learning intentions. Do not be afraid to use picture books written for much younger children with older classes, making it clear that they would have read them some years ago to avoid any sense that they are using babyish resources!

- Nick Butterworth's stories about Percy the Park Keeper – such as *After the Storm* (1992) – are ideal for this activity; as we read the stories, we gradually find out more about the park. I know it has a pine wood, a stream, a lake, for example, and mapping out an idea of what it looks like would provide the basis for some journeys work as Percy trudges round the paths to attend to what needs doing.
- Julia Donaldson's *What the Ladybird Heard* (2009) is a super book for supporting children in understanding maps and routes. I will not give the game away; you need to read it for yourself!
- Janet and Allan Ahlberg's *The Jolly Postman* (1986) is a timeless classic and lends itself to mapping out the journey the postman takes on his bicycle.

- Val Biro's Gumdrop stories – such as *Gumdrop and the Elephant* (1980) – describe the adventures of a vintage car. Any story involving a mode of transport has great potential for this topic.

What other stories would you want to add to the list?

Pupil activity 11.4

How do I get to where I want to go?
Y1 to Y4

Learning objective: Ability to give and receive directions.

When the focus is on direction rather than position, play mats are a useful resource, or a town layout could be set up in the classroom or drawn on the playground. Having set up your 'town', prepare cards with all the places shown on the map. Children pick two place cards plus a third card with an age-appropriate directional term; they direct a partner how to get from one place to the other using, amongst other words, the term on the third card. Incidentally, this also works well in a foreign language when learning how to give and follow instructions!

Pupil activity 11.5

Treasure hunt
Y2 to Y4

Learning objective: Ability to identify position using early coordinates.

Using and creating maps where the squares are labelled using letters and numbers is the easiest way to introduce and practise the idea of basic coordinates, for example, to position the pirate's treasure! Once maps like this one have been designed, use them to play games. For example, the 'pirate' secretly writes down where the treasure has been buried and the players take it in turns to roll two dice: one a regular dice, the other labelled with the letters A to F. They place their counter on the appropriate square and play continues until the treasure has been unearthed. The children take it in turns to be the pirate.

The game 'Battleships' offers a slightly more advanced idea for use with older children, with ships of different shapes and sizes being secretly positioned over several squares and the opponent attempting to sink them by guessing the coordinates.

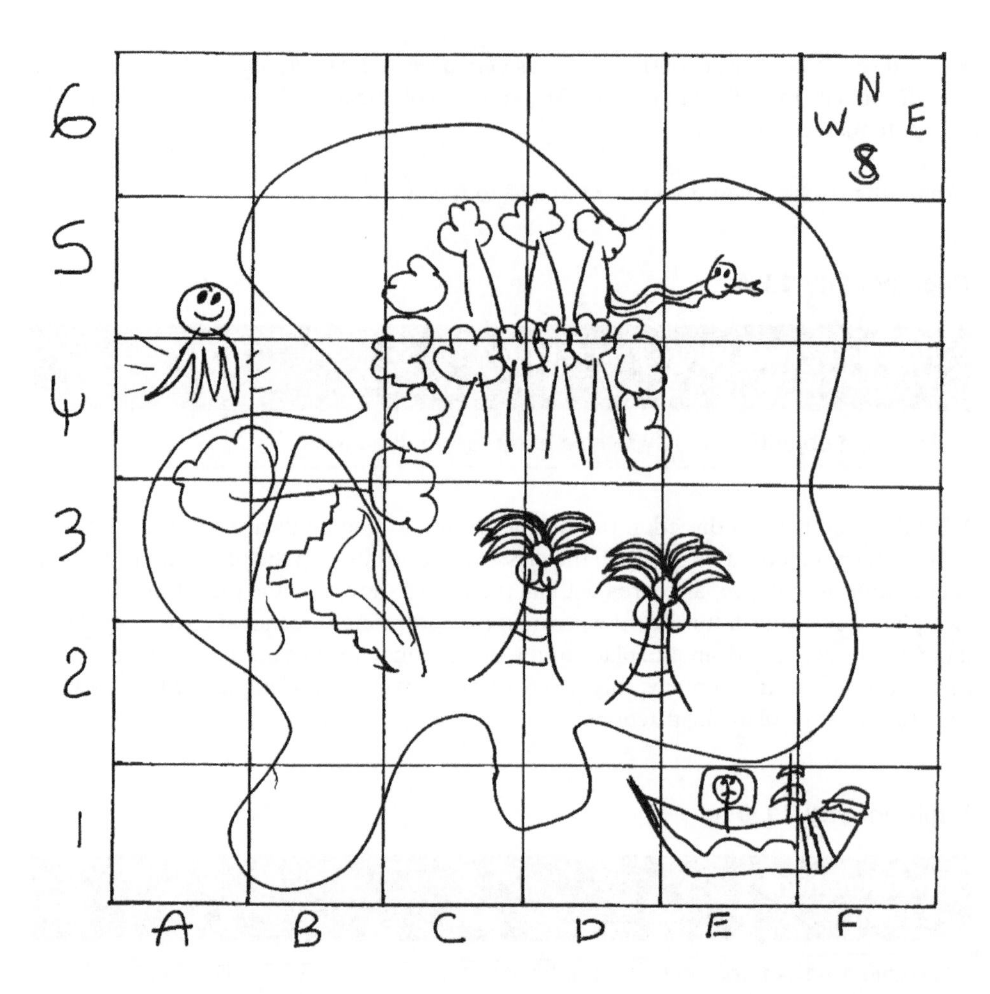

Pupil activity 11.6

What are coordinates?
Y5 to Y6

Learning objective: To begin to appreciate the role numbers can play in specifying position.

Peg boards can be used to introduce the idea of Cartesian coordinates. Set up a scenario where a child secretly chooses where to put a peg in their board and then has to explain to a friend (as if on the phone) where they have to put their peg so that the two boards match. This can be done several times using different coloured pegs and it's likely that the children will use numbers in some way to help describe the hole that the peg needs to go in. Following this activity, the introduction of conventional coordinates becomes a logical progression. It may help to mark one corner of the board for clarity, with this eventually becoming related to the origin.

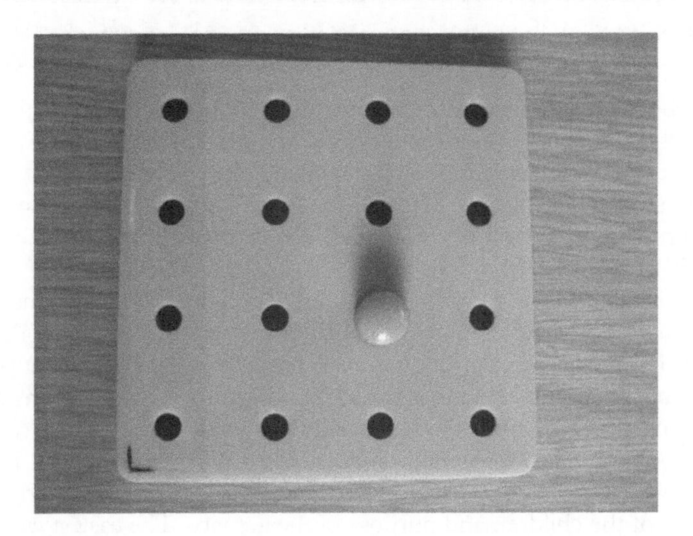

Pupil activity 11.7

Interpreting coordinates
Y5 to Y6

Learning objective: To understand the relationship between plotted points and shapes made by joining them.

Coordinates offer a great opportunity to make links with shape work through predicting the shapes that will be generated if you were to plot and join up particular pairs of coordinates. In order to get children really thinking, try to throw in a few red herrings – are there any here? Encourage the children to visualise plotting the points and joining them before testing any out.

- (3,4) (5,2) (5,4)
- (1,2) (3,1) (5,0)
- (1,2) (3,1) (5,0) (4,2)

Obviously as well as you providing sets of coordinates for the children to think about, they can also make up and check sets for their friends.

Pupil activity 11.8

Positions please!
Y1 to Y6

Learning objective: Ability to reason using positional language.

As well as rehearsing language associated with position, this type of activity also provides wonderful reasoning opportunities in solving the problem of where the items go. Whereas the example below couples the language of position with that of shape, and would be most suitable for Key Stage 2, the focus could purely be upon positional language with younger children, using numbers, or items linked to other class topics such as 'fruit' or 'vehicles'.

The idea is that a class or group of children negotiate where the items should be in the grid to satisfy all the (written or verbal) clues they have been given. Revealing one clue at a time is a good way of promoting reasoning; thinking about what you know for sure versus 'maybes'. You could also ask a child or group to take responsibility for the position of a particular shape. Whilst I usually provide cut-out shapes (or whatever), there would be nothing to stop children drawing or making their own. I have always found a 3×3 grid is ideal, but you could of course use a bigger or smaller grid, depending on the age of the children and purpose of the activity. The easiest way to generate the clues is to put things in the grid at the planning stage, write the clues ensuring you incorporate the range of language you wish to cover, and finally check that the clues given are sufficient to complete the task. You may wish to have items which slot into position in the grid by default (like the hexagon in the example pictured) or to include 'redundant information' which does not help to solve the problem, for example 'Four of the shapes contain right angles'. As a variation on the teachers creating the clues for completion of the grid, the children could of course make up their own.

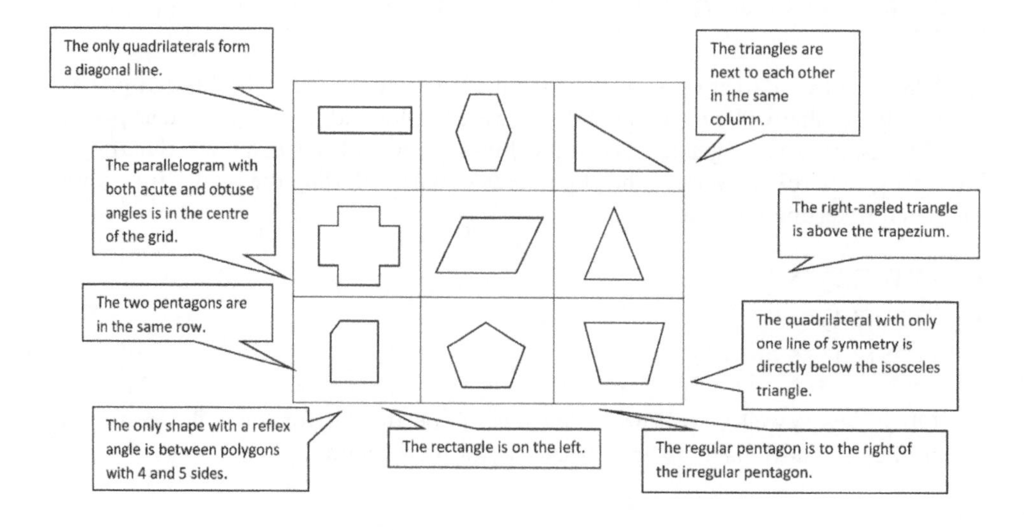

Pupil activity 11.9

The fidgety ladybird
Y2 to Y6

Learning objective: Ability to describe changes of position using the language of rotation and translation.

To link with another topic, my class once studied pictures of pairs of ladybirds in different positions, describing the movement from one to the other in terms of rotation and translation. The children went on to create a whole row of fidgety ladybirds with consistent repetition of the same movements for others to identify. Note that such pictures can also be used to explore whether a line of symmetry exists between the two ladybirds (possibly in this case) thus reflective symmetry could be the focus. This can be checked using a mirror.

An option for older children is to begin to relate this to describing the movements using a coordinate grid. As well as transformations, children will be rehearsing their coordinates knowledge, for example by describing the head of the ladybird moving from one coordinate to another through a rotation of however many degrees about a certain point.

Pupil activity 11.10

Out for a spin
Y1 to Y4

Learning objective: To appreciate the different ways in which directions and size of turn can be described.

Depending on the age of the children, turning movements (both size and direction) tend to be described in different ways. Dealing practically with magnitude helps to reinforce that angle is a measure of turn and is by far the best way of helping children get to grips with the relationship between the terminology and the magnitude and direction of the turn. With everybody standing facing the same direction, instructions of the following type are given in a game akin to 'Simon says'. If you give only one part of the instruction (the direction or the size), the children should stay still. This helps to draw their attention to the fact that there are two pieces of information:

- Quarter turn to face the window
- $\frac{3}{4}$ turn, turning to the left
- Half-turn clockwise
- 90° turn anti-clockwise
- 170° turn [Almost 180° so almost a $\frac{1}{2}$-turn but no direction was given so no turn!]

Note that these start simple and gradually use more complex styles of instruction to illustrate differences across year groups, but with all age groups vary the language from time to time and make sure you include the occasional full turn. Also emphasise, through your choices, that turns of the same size in the opposite direction can 'undo' a previous movement and that successive turns can complete a full turn, such as $\frac{1}{4}$ then $\frac{3}{4}$ turns clockwise.

Pupil activity 11.11

Transformation teacher
Y4 to Y6

Learning objective: Ability to explain the different types of transformation.

The idea with this activity is that the children work as a team to make a short video explaining one aspect of transformation; perhaps one chosen at random from a set of cards to ensure that different groups cover different aspects. Give each group the opportunity to watch the rehearsal of another group; this way every group receives what is hopefully constructive criticism before they produce their final video.

Pupil activity 11.12

Compass points
Y3 to Y4

Learning objective: To develop familiarity with the four or eight points of the compass.

Having established roughly which direction north is in and fixed the other points of the compass in relation to it, the simplest way of reinforcing compass directions is to incorporate them into an activity similar to activity 11.10. The focus switches to facing north or taking three steps to the east or whatever.

Treasure maps (see activity 11.5) also lend themselves to a NSEW focus, and the game needs cards or spinners prepared with the four or eight compass points. This time agree with the other players (maximum four) where the treasure is buried. Each player starts at a corner of the map and a dice roll tells you how many squares you can move that turn. Two directions cards (or spins of the spinner) give your possible directions of travel. If your number of steps and/or direction forces you off the map, you return to your start position. Who can get to the treasure first?

Pupil activity 11.13

Roamer
Y2 to Y6

Learning objective: To develop spatial ability relating to estimation of distance and angle.

Many schools have programmable robots available for classes to borrow, and these can be used for rehearsal of estimation skills relating to distance and angle. A Roamer's default setting is to work in degrees, but it can be set to turn in numbers of right angles. Even quite young children can be told that 90 is a special number which results in a right-angled turn and will happily explore bigger and smaller numbers to discover how far the robot turns. Setting up some sort of obstacle course then tests the children's ability to estimate the distances and angles required. Children should be encouraged to articulate what they plan to input as an opportunity for them to use appropriate language.

Pupil activity 11.14

Logo
Y3 to Y6

Learning objective: To understand that, to draw a shape, angle of turn is not necessarily the same as the interior angle of the shape.

Logo software (available from MSW Logo, http://www.softronix.com/logo.html) provides a really nice opportunity for children to engage in some basic computer programming. My approach is usually to give the children the basics required for drawing something like a square (or to at least check that they have this knowledge) and to then give them a little freedom to explore the making of other shapes through which they will need to investigate turns of different sizes. Access to paper is helpful for working out the angles since the turns required may not be what learners expect (see diagram) and it can help to be able to physically turn the page.

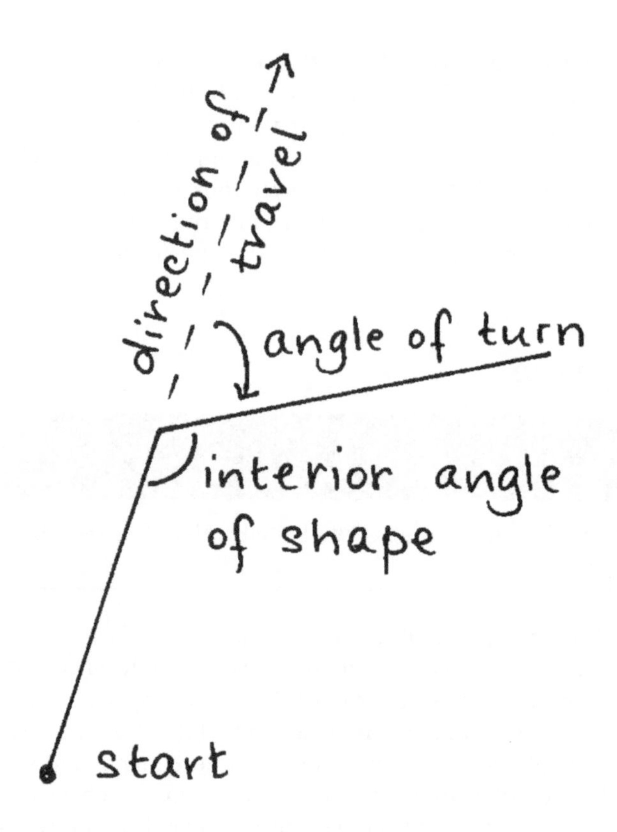

Children can choose their favourite to print and describe, encouraging the use of specific age-appropriate language, such as making reference to different types of angle.

Pupil activity 11.15

Exploring lines of symmetry
Y2 to Y5

Learning objective: To develop understanding of what constitutes a line of symmetry.

Open-ended tasks work well for exploring lines of symmetry, assuming children have access to a rich range of shapes or symmetrical images to investigate. A circle may not seem an obvious choice, but actually it's great if children are able to demonstrate knowledge that circles potentially have an infinite number of lines of symmetry! Things like letters of the alphabet are also rich for exploration with some letters, such as 'Z' and 'N', proving more challenging than others. I have quite a number of letters with lines of symmetry in my name; what about you?

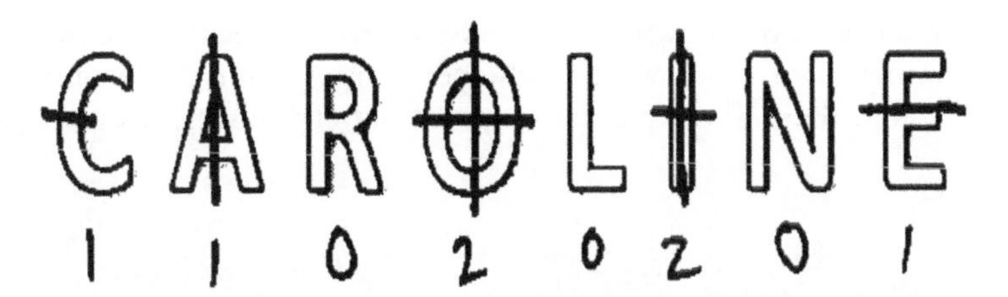

Working collaboratively, children will challenge each other about answers they disagree with and these can then be shared with the class in an attempt to resolve the disagreement.

Pupil activity 11.16

Reflective symmetry
Y3 to Y6

Learning objective: To develop understanding of the relationship between symmetry and reflection.

Rather than lines of symmetry running through a shape, here we explore the way in which patterns can be reflected across a line. Pattern blocks are ideal for this as they can be flipped over to show what they would look like on the other side of the line; alternatively, a mirror can be used to check.

As well as pattern blocks, computer software such as that found on the TES iboard website (http://www.iboard.co.uk) can also offer ways of investigating reflective symmetry, as does squared paper. Horizontal and vertical lines of symmetry prove the easiest, but here we see a diagonal line.

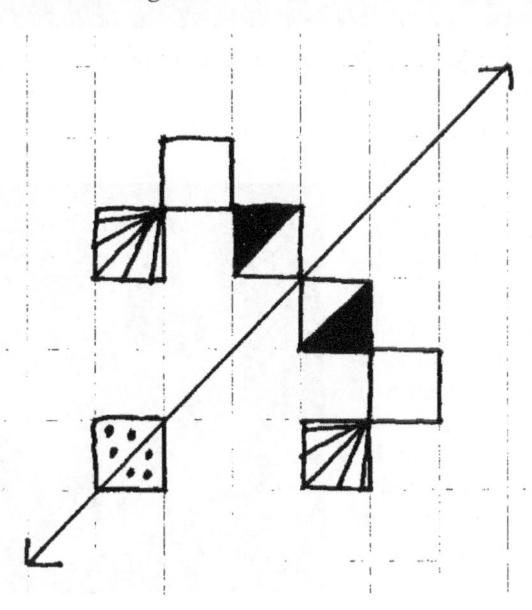

Children can also be shown completed patterns to decide whether they have reflective symmetry; in other words, whether a line or lines of symmetry could be drawn through them. This picture made of pattern blocks has just one line of symmetry.

Pupil activity 11.17

Rotational symmetry
Y4 to Y6

Learning objective: Ability to identify order of rotational symmetry.

As well as presenting children with photographs like this, ideal for working out the order of rotational symmetry, children can also be encouraged to take and categorise their own pictures. Discussion with and between children will probably involve turning the pictures to check for rotational symmetry. It may help to cut round the key part of the picture so the corners of the photos do not confuse; it can also help to mark a point somewhere on the edge so that you know when to stop turning!

Summary of key learning points:

- mathematics can help us to describe and quantify position, direction and movement;
- language is integral to spatial topics;
- links with real-life and geography topics will be mutually beneficial in strengthening children's understanding;
- careful visual discrimination will be required to identify aspects such as symmetry;
- much learning related to space is best served by practical experience rather than use of the interactive whiteboard.

Self-test

Question 1

In activity 11.7 the second shape will be a triangle and the third shape a quadrilateral. True or false?

Question 2

Thinking about enlargement, if I make each side of my square three times longer, the area will be nine times its original size. True or false?

Question 3

ESE is the compass point which occurs between east and south-east. True or false?

Question 4

The same result can be obtained by combining different types of transformation. True or false?

Self-test answers

Q1: (False) If you try drawing either of these you will discover that three of the coordinate pairs generate a straight line, thus the second set draws only a line, and the third set will result in a triangle in spite of there being four pairs of coordinates.

Q2: (True) Growing the little square, say 1 cm², by a scale factor of 3 will indeed result in a square of 9 cm². This is because each length is trebled (1 cm becomes 3 cm) so the resulting square would be 3 cm × 3 cm = 9 cm². If you remain unconvinced, you might wish to start with a square with different length sides to check that the same thing happens: if side lengths are increased by a scale factor of 3, the area will be 9 times larger.

Q3: (True) ESE is indeed the compass direction between E and SE. The points beyond the first eight are always named according to the points either side, with the major one stated first.

Q4: (True) Whilst one child might interpret one image as a reflection of another, a different child could describe this in terms of rotation and translation; both would be right.

Misconceptions

'As the wind is going south it is called a southerly wind'

'I can plot all my coordinates and then join them up in whatever order I like as it won't make any difference'

'Reflection means you have to repeat one side of the page on the other side'

'As the lines are twice as long the area must be twice the size too'

'I have cut this shape in half so it must be symmetrical'

12

Conclusion

The importance of mathematics tends not to be disputed; hopefully my passion for the subject has helped to enthuse you in some small way. But good teaching of it is vital, especially as we know that so many people do not share a love of mathematics. So whether you have read the book from cover to cover, or dipped in and out of chapters on particular topics, you have hopefully been encouraged to think about aspects of your mathematics teaching a little more deeply. If you have found yourself considering some things which you had not really thought about before, then that will be beneficial. It mirrors what should be happening in our mathematics classrooms; the children should be prompted to think hard about things they have not thought about before. Whilst it might be difficult to identify when in my life I might use some of the mathematical content I learn, if mathematics can teach me powerful reasoning skills, then I have certainly benefitted from my mathematics education.

Authors sometimes write about learners ideally having a 'feel' for number (Haylock, 2010). For instance, the numbers 16 and 17 sit side-by-side yet 'feel' different; whilst 16 is a multiple of 2 and 4 and 8, 17 is a prime number. But mathematics encompasses more than just number, and we want children to develop a curiosity about the worlds of number, shape, and so on. This curiosity happens in classrooms where teachers are excited about mathematics and present interesting opportunities to entice the children to explore mathematics for themselves. Effective teachers support children in having the confidence to investigate mathematical topics without fear of failure. I do not mean that mathematicians do not fail; they do. To succeed in mathematics it has to become OK to get things wrong and to say 'I don't know' when asked a question.

Mathematicians also talk to other mathematicians; they share ideas and argue about things. Do the children in your class and your school have such opportunities? Do they scribble unintelligible jottings when they are grappling with mathematics, such as a challenging problem or tough calculation? Or is all their recording about filling in answers on worksheets and doing so neatly? Once the mathematics work is completed, who judges it to be right or wrong? Ideally the child will be given some responsibility for this ('I'm sure I'm right because...'), but sadly it seems to be rarely the case. Aim to allow your class the freedom to begin to think and behave like mathematicians. Incidentally, the notion of freedom will also pay dividends in terms of differentiation. If children are able to make choices about what they work on, they tend to get it about right most of the time; this helps to remove the 'false ceilings' that so many children have imposed upon them when a teacher dictates particular work for particular groups.

Hopefully you have found some new ideas to try in your own classroom and have been convinced that good mathematics teaching can come from simple ideas and from adapting existing ideas or from using materials more creatively. With active professional development in mind, one suggestion would be to now work with a colleague: discuss one of the chapters; agree to try out some of the ideas in your own classrooms; commit the time to come back together to discuss how it went. But whether you are able to do this or are working alone, all that remains for me to say is: enjoy your mathematics teaching!

References

Ahlberg, J. and Ahlberg, A. (1986) *The Jolly Postman*. London: Heinemann.

Andrews, P. and Sayers, J. (2003) Algebraic infants, *Mathematics Teaching*, 182: 18–22.

Askew, M., Brown, M., Rhodes, V., Wiliam, D. and Johnson, D. (1997) *Effective Teachers of Numeracy: Report of a Study Carried out for the Teacher Training Agency*. London: King's College, University of London.

Barmby, P., Bilsborough, L., Harries, T. and Higgins, S. (2009) *Primary Mathematics: Teaching for Understanding*. Maidenhead: McGraw-Hill/Open University Press.

Biro, V. (1980) *Gumdrop and the Elephant*. London: Hodder Children's Books.

Boaler, J. (2009) *The Elephant in the Classroom: Helping Children Learn and Love Maths*. London: Souvenir Press.

Butterworth, N. (1992) *After the Storm*. London: HarperCollins.

Cockburn, A. and Littler, G. (eds) (2008) *Mathematical Misconceptions*. Los Angeles and London: Sage.

Cooke, H. (2001) *Passport to Professional Numeracy*. London: David Fulton.

DfEE (1999) *The National Numeracy Strategy: Framework for Teaching Mathematics*. London: DfEE Publications.

DfES (2006) *Primary Framework for Literacy and Mathematics*. London: DfES Publications.

Donaldson, G., Field, J., Harries, D., Tope, C. and Taylor, H. (2012) *Becoming a Primary Mathematics Specialist Teacher*. Abingdon: Routledge.

Donaldson, J. (2009) *What the Ladybird Heard*. London: Macmillan.

Faux, G. (1998) Gattegno charts, *Mathematics Teaching*, 163.

Gelman, R. and Gallistel, C. (1978) *The Child's Understanding of Number*. Cambridge, MA: Harvard University Press.

Gravett, E. (2008) *The Odd Egg*. London: Macmillan.

Haylock, D. (2010) *Mathematics Explained for Primary Teachers* (4th edition). London: Sage.

Haylock, D. and Cockburn, A. (2008) *Understanding Mathematics for Young Children: A Guide for Foundation Stage and Lower Primary Teachers*. Los Angeles: Sage.

Hewitt, D. (2009) From birth to beginning school. In J. Houssart and J. Mason (eds), *Listening Counts: Listening to Young Learners of Mathematics*, pp. 1–15. Stoke-on-Trent: Trentham Books.

Hutchins, P. (1986) *The Doorbell Rang*. London: Bodley Head.

Knee, K. (2010) Discussion in the mathematics classroom. Unpublished independent project available in the University of Chichester library, Bognor Regis.

Mooney, C., Ferrie, L., Fox, S., Hansen, A. and Wrathmell, R. (2012) *Primary Mathematics: Knowledge and Understanding* (6th edition). London: Learning Matters.

Orton, A. (ed.) (2005) *Pattern in the Teaching and Learning of Mathematics*. London: Continuum.

Pinczes, E.J. (1995) *Remainder of One*. Boston: Houghton Mifflin.

Potter, L. (2006) *Mathematics Minus Fear*. London: Marion Boyars.

Rickard, C. (2010) Algebra homework, *Mathematics Teaching*, 219: 14.

Rowland, T., Turner, F., Thwaites, A. and Huckstep, P. (2009) *Developing Primary Mathematics Teaching: Reflecting on Practice with the Knowledge Quartet*. Los Angeles and London: Sage.

Ryan, J. and Williams, J. (2007) *Children's Mathematics 4–15: Learning from Errors and Misconceptions*. Maidenhead: Open University Press.

Suggate, J., Davis, A. and Goulding, M. (2010) *Mathematical Knowledge for Primary Teachers* (4th edition). Abingdon: Routledge.

Thompson, I. (2009) Place value? *Mathematics Teaching*, 215: 4–5.

Index

RETHINKING SUPERHERO AND WEAPON PLAY

Superheroes and Children's Moral Development

Steven Popper

9780335247066 (Paperback)
April 2013

eBook also available

This book explores children's war, weapon and superhero play with a view to examining its potential (positive) impact on developing moral values and sensibilities, and to the many moral themes available for children's exploration during their engagement with such play, and the traditional and continuing need for children to receive a good moral education, with reference to many ideas from educational philosophy.

Key features:

- It links examples of children's real-life play and perspectives to theories about play, moral development and narrative psychology
- It explores the continuing attraction of classical dualism (i.e. good versus evil) for children and various educational perspectives about this
- Contains a wealth of learning opportunities and suggestions of ways to use superheroes to advance children's moral, philosophical and emotional thinking

www.openup.co.uk